thirteen DOOMS
how humanity ends

Word Balloon Books!

thirteen DOOMS

how humanity ends

by

James Maxey

Copyright © 2024 by James Maxey
ALL RIGHTS RESERVED

Cover by James Maxey

Interior illustrations by Albrecht Dürer and are in the public domain

The author may be contacted at
james@jamesmaxey.net

ISBN: 9798308344193

*For Calvin Powers
and other shadows who stretch before me.*

TABLE OF CONTENTS

Doomed 1
Thirteen Horsemen 7
The First Horseman: Famine! 13
The Second Horseman: Comets! 27
The Third Horseman: Overpopulation! 37
The Fourth Horseman: Climate Change! 49
The Fifth Horseman: War! 59
The Sixth Horseman: Trash! 67
The Seventh Horseman: Aliens! 75
The Eighth Horseman: Plague! 85
The Ninth Horseman: Artificial Intelligence! ... 93
The Tenth Horseman: Natural Stupidity! 105
The Eleventh Horseman: Supervolcanoes! 115
The Twelfth Horseman: Transfiguration! 123
The Thirteenth Horseman: Entropy! 133
Welcome to the Human Condition 139
Afterword The Reserve Horsemen 143
About the Author 146
Books by James Maxey 147
Links 149

All go unto one place;
all are of the dust,
and all turn to dust again.

Ecclesiastes 3:20

DOOMED

Imagine a world in the midst of a terrible apocalypse. Over 95% of humankind has already perished. The scope of the destruction is so immense that whole nations have vanished. The overwhelming majority of the dead have no one to mourn them. The people who knew the names of these lost souls have also passed away. For most of the deceased, we don't know the location of their graves.

The Grim Reaper hasn't put away his scythe. All over the world, people are sick and dying. Thousands perish every hour. In the time it takes to read this page, hundreds will lose their lives. It's nearly guaranteed someone's going to keel over before you even finish this sentence.[1]

Despite the horrifying death toll, in the midst of the ongoing destruction, people cling to hope. Perhaps they'll be spared. Perhaps the people they love won't be touched by the tragedy. These people, alas, are deluded. In the end, death will claim them, their friends, their families, and their little dogs, too.

The first word of this book was "imagine." That was a buffer, a little padding to soften the harsh truth. You don't need to imagine this world of unstoppable death.

---

[1] It's not the book's fault. Probably.

It's our world. We're living in it, but not for long. Individually and collectively, our time is running out.

On a planetary scale, humans arose on Earth in relatively recent times. The available evidence points to our species being roughly a quarter of a million years old. That might sound ancient, but it's less than .01% of the age of the planet. The world did just fine for a very long time without us. Once humans did show up, our ancestors made up for lost time by spawning roughly 120 billion descendants. Despite furious efforts at forestalling their mortality through hard work, careful planning, and appeals to numerous divinities, 112 billion of these distant relatives of ours are now dead.

We've been born into a world in the throes of a protracted apocalypse. Nothing will save us.

We're doomed.

The only questions remaining are when we'll meet our ends, and how.

The most likely answer is that it will be some relatively common, mundane cause that carries us out of this life. Old age, disease, and accidents are the big killers. Some of us may meet our end due to war. Murder is also a possibility. Some of us will get tired of waiting and take our own lives. Whatever the cause, death will claim us one by one.

Unless, of course, we all die collectively, in one fell swoop. Extinction is part of the natural order. You probably are familiar with a few famous extinct species, like the dinosaurs, the dodo, or the Tasmanian tiger. Our nearest related species have an especially poor track record. All the upright, big-brained, tool-using primates related to *Homo sapiens* have vanished from the planet.[2] Some of the better-known members of this group are the Neanderthals, *Australopithecines*, and *Homo floresiensis*. But the extinctions I can name aren't even the tip of the iceberg. They're more like a snowflake sitting atop that iceberg, with an impossibly large mass of vanished species underneath that are forever lost, and will remain unknown.

How do species meet their end? Climate change has wiped out numerous species. Sometimes big rocks come flying out of space and wipe out multiple species at once in a planetary cataclysm. Not that our planet needs outside help to eliminate species. There have been volcanic eruptions so massive that they've been just as destructive, if not more so, than the collision of a comet.

Disease has also taken a toll. Odds are, you haven't sat in the shade of any American Chestnut trees lately.

Perhaps the darkest driver of extinction is, um, us. People, I mean. Humans have driven countless species to extinction. When people first reached North America, they needed only a few thousand years to wipe out a whole catalog of megafauna, like giant sloths, mastodons, and dire wolves. The Neanderthals I mentioned a few paragraphs back? Odds are good that we *Homo sapiens* helped usher them off the planet.

What will be the next species we extinguish?

One likely candidate species that mankind might

---

[2] This is one of many bits of evidence that our brains cause more trouble than they're worth.

erase from the planet is... *Homo sapiens*. I was keenly aware of this growing up, in the fading years of the Cold War. Then, popular culture delivered a non-stop flood of movies and books about nuclear annihilation. The Cold War mostly vanished with the breakup of the Soviet Union, but there are still enough bombs that, if we do start World War III, there might not be anyone left to fight in a World War IV.

Somehow, despite the continued existence of this literal doomsday arsenal, today the threat of nuclear war feels almost antiquated compared to other dangers.

Climate change in previous eras has wiped out numerous species. Today, we're unintentionally engineering a massive alteration of our atmosphere. It's an open question whether we'll survive the changes we've already set in motion.

Diseases have devastated human populations in previous eras, and we've reached a level of technology where we no longer need to worry that random mutations of naturally occurring viruses might produce some devastating plague. There are labs where humans add and subtract genes to viruses, creating lethal microbes that can spread around the globe in a matter of weeks.

But an extinction that unfolds over weeks might be a very slow way to go. We could go extinct nearly overnight if our technology runs amok. We're in the early years of dabbling with artificial intelligence. So far, the machines we've built are limited in power, and relatively obedient. But, will there be a tipping point? Is it possible we'll create an intelligent machine that creates an even more intelligent machine with a mind so far beyond ours that the machines no longer need us, and actually find us inconvenient? How long do you think you'd survive in a world where killer drones are hunting door to door, determined to wipe out the last human?

As long as we've entered the science fiction realm of killer robots, what about the other classic threat, the alien invader? Space is vast, and one can plausibly argue that the skies house millions of civilizations more advanced than our own. They might be on their way here right now, armed with weapons of conquest we can't even imagine.

And we don't need imagination at all to contemplate other threats from outer space. Our planet is whirling through a universe filled with massive space rocks. Asteroids, comets, and even small planets have collided with us before, and it's only a matter of time before we're in the crosshairs of another planet-breaker.

There's an element of mercy in the swift end. A bleaker possibility is that we might die off slowly, over centuries, as we exhaust the resources of our planet. The last starving humans might dig vainly for roots hidden beneath parched dust, beneath the dim light of a sun eternally shrouded by an atmosphere thick with poisons.

I'm sorry if these possibilities are making you anxious. Try to remain calm. Panic won't help!

On the other hand, it probably won't hurt. A little fear can be a good thing. If the dodos had been a little more afraid of the strange apes turning up on the shore of their island, they might still be around.

In the pages ahead, we'll take a closer look at our potential dooms. We can contemplate them coolly, at a distance, before facing them in a time of crisis. This gives us a chance to access all the potential extinction level threats and form strategies to deal with them. Perhaps there's reason for hope that, by acting collectively or as individuals, we might yet save our species.

THIRTEEN HORSEMEN

Plague! War! Famine! Death! These are the classic "Four Horsemen of the Apocalypse." It's not a terrible list of dooms, especially if you were making your list in the year 95 AD and had never heard terms like "climate change" or "nuclear winter." Of course, humankind's collective knowledge has grown quite a bit in the last twenty centuries. Today, with our increased understanding of the workings of the universe, do we still need to worry about these four atavistic horsemen of destruction?

Yes. And now the original horsemen have been joined by fresh cavalry. Behold, the clouds shall part to reveal the modern horsemen, the veritable legion of dooms we face. Some of these fates aren't our fault. Many dooms are simply woven into the fabric of nature. But these natural dangers have been joined by a host of hazards that are the product of man's carelessness, arrogance, and greed.[3]

In the course of researching this book, I was able to identify thirteen potential pathways to human extinction[4]. For an overview of the dangers we face, here's my updated list of the Horsemen of the

---

[3] Also, stupidity. Mostly stupidity.

[4] I can actually name, like, thirty but thirteen is a spooky number, and I'm too lazy to do thirty chapters.

Apocalypse. Listen! You can hear their hoofbeats!

1. Famine! We'll begin our slouch toward doomsday by taking a close look at the OG Apocalypse. Is Famine caused simply by a poor harvest? Or might sinners be to blame?

2. Comets! Comets and asteroids have a track record of creating mass extinctions. Something big is on a collision course with us. Can we do better than the dinosaurs?

3. Overpopulation! Birds do it, bees do it, and the teeming masses of humanity do it at an alarming rate. Will overpopulation lead to an empty planet?

4. Climate change! CO_2 and other greenhouse gasses are transforming our atmosphere at an accelerating rate. Can we survive the coming heat age?

5. War! More than one nation possesses an arsenal capable of unleashing a world-wide firestorm. How much longer can we hold off this destruction?

6. Trash! Will we eventually go down in a literal sea of trash?

7. Aliens! Mars probably doesn't need our women, but that doesn't mean we aren't a good candidate for an alien invasion. The universe is likely filled with civilizations more advanced than our own.

If they want to invade, do we have any hope of stopping them?

8. Plague! Another classic horseman of doom. Can our interconnected world survive a plague even more gruesome than the Black Death?

9. Artificial Intelligence! Will digital minds inherit the Earth?

10. Natural Stupidity! Can our complex world endure a rising tide of simpletons?

11. Super Volcanoes! The geysers of Yellowstone are more than just a tourist attraction. They're also a harbinger of our inevitable doom.

12. Transfiguration! Will we willingly end the human race in order to become something better?

13. Entropy! Alas, thermodynamics.

Of course, the Thirteen Horsemen, while fearsome, have a lot of work cut out for them. Humankind has so far been able to adapt to an impressive range of hardships and challenges, and this was long before we had access to modern technology.

We survived the worst of the Ice Age, when pretty much all land north of the Mason Dixon line was frozen under ice up to two miles thick. In fact, we didn't just survive the Ice Age. Humans spread their range, trekking out of Africa and adapting to live on every continent except Antarctica.[5] Then, as the ice sheets covering the northern hemisphere retreated, humans, who likely lived predominantly near water, managed to carry on through 400 feet of sea level rise.

Given that we survived these terrifying catastrophes at a time when human technology consisted of little more than sharp rocks and campfires, our modern technology should allow us even more tools to hold the Horsemen of the Apocalypse at bay.[6] It's possible, or even likely, that some of these dooms unfold exactly as predicted and we'll simply muddle through as a species, battered and bruised, but still alive.

On the other hand, some of these horsemen might be so good at their jobs that they not only kill every living human, but leave in their wake a dead, sterilized Earth.

With this in mind, we'll be judging each of these potential apocalypses on their ability to trigger four different outcomes:

1. Should I panic? Will this kill me?

2. Could this decimate us?

3. Could this send us back to the Stone Age?

4. Will this cause extinction?

---

[5] And you should look up information about the Piri Reis map if you want to consider the possibility that ancient people might have at least reached that continent.

[6] Unless modern technology destroys us, which is a serious possibility.

A catalogue of dooms we can't possibly avoid would be depressing. I genuinely feel it's my duty as an author, and as a fellow resident of the planet, to offer potential solutions to the problems we face. So the final question we'll examine in each chapter is: *Is there hope?*

Because, in nearly every case, there is hope. Keep in mind that I'm a science fiction author. While I read widely, both fiction and non-fiction, I'm not an academic immersed in the deep study of these subjects. But, we live an era where there are people way, way smarter than me[7] devoting their lives to thwarting the horsemen riding toward us.

With this in mind, let's plunge ahead, and meet the First Horseman.

---

[7] People who put actual references in footnotes instead of jokes.

And I will show you something different from either
Your shadow at morning striding behind you
Or your shadow at evening rising to meet you;
I will show you fear in a handful of dust.

T.S. Elliot, *The Waste Land*

THE FIRST HORSEMAN: FAMINE!

Let's begin our journey by opening our Bibles. Don't worry. You haven't stumbled into a Sunday School lesson. But as we dive into our examination of world-shattering events, it's worth looking at why the word "apocalypse" is in our everyday vocabulary.

I don't think the word "everyday" is an exaggeration. The apocalypse is a robust thread in the fabric of modern popular culture. Books set in post-apocalyptic worlds routinely show up on bestseller lists. Do a quick search on Amazon music, and you'll find tens of thousands of songs that contain "apocalypse" in the title. Some summers, it feels like every third movie has the heroes racing to prevent the apocalypse. Even cartoons aimed at kids romp in the apocalyptic playground. *Adventure Time,* for instance, is set on a shattered Earth in the aftermath of the Great Mushroom War.

The apocalypse isn't some esoteric idea that only a handful of nerds care about. The idea of the apocalypse is built into the very foundation of civilization.

How did the word "apocalypse" make it into our language? Today, most people understand that an apocalypse is an event involving terrific destruction. But, the original meaning of "apocalypse" had nothing to do with destruction. The original meaning was "to uncover," or "to reveal."

It was in the sense of "revealing" that the word became famous. The last book of the Bible, Revelation, begins in the original Greek with the word "apokalypsis." Translated, the full beginning reads, "This is the revelation of Jesus Christ, which God gave Him to show His servants what must soon come to pass." The apocalypse was information about how the world would end, not the ending itself.

But, what an ending! The book of Revelation is where the Four Horsemen of the Apocalypse ride forward on Judgment Day. The horsemen are pretty metal. The passage where they're introduced reads like an angsty fourteen-year-old's daydream.[8]

"I heard one of the four living creatures saying with a voice like thunder, 'Come and see.' And I looked, and behold, a white horse. He who sat on it had a bow; and a crown was given to him, and he went out conquering and to conquer." Then the second horseman shows up, on a red horse, and he's got a terrifying sword. The third horseman rides in on a black horse, and he's carrying scales. He immediately starts complaining about the price of grain.[9]

---

[8] Fourteen-year-olds at the time of Revelation were going through puberty and a midlife crisis at the same time.

[9] "A quart of wheat for a denarius, and three quarts of barley for a denarius." What was a denarius? The Oxford English Dictionary says it was "ancient Roman silver coin, originally worth ten asses." Essentially, the third horseman was shouting, "Your asses can't buy food!"

Thirteen Dooms

But the first three horsemen are just the warm up act. The real star shows up next: "…and behold, a pale horse. And the name of him who sat on it was Death, and Hades followed with him."

By the end of the next verse, a quarter of the Earth's population is dead. Then, things get worse! Next up are a bunch of disasters all happening at once: "…behold, there was a great earthquake; and the sun became black as sackcloth of hair, and the moon became like blood. And the stars of heaven fell to the earth, as a fig tree drops its late figs when it is shaken by a mighty wind. Then the sky receded as a scroll when it is rolled up, and every mountain and island was moved out of its place."

Then angels show up! "So the seven angels who had the seven trumpets prepared themselves to sound. The first angel sounded: And hail and fire followed, mingled with blood, and they were thrown to the earth. And a third of the trees were burned up, and all green grass was burned up."

There are still six angels to go, each heralding disaster after disaster. By the time it's all over, two thirds of mankind are dead. There aren't as many trees, the fishing is terrible, and the mountains and islands are scattered willy-nilly.

Now that's an apocalypse!

In fact, it's *the* Apocalypse! And it's not just some old story. Today, there are millions of people around the world who still believe[10] this is how the world will end.

---

[10] Or even hope.

Taking some of the sting out of the vision is that it's been 2000 years since John said that this doom "must soon come to pass." The word "soon" has some wiggle room built in, but after twenty centuries, a reasonable person might conclude that John has a few details wrong.

However, this chapter isn't about what's wrong with John's vision of doomsday. Instead, we'll look at something he got right. We need to take a closer look at the Third Horseman. It's worth contemplating why, among riders leading the legions of Hell, a dude shouting out the price of grain struck John of Patmos as a figure of terror.

Revelation is a warning that God is so upset about sinners that he's going to kill off three quarters of the globe. Is there some hidden connection between sin and money we should look at?

The Bible has several admonitions about money. Two well-known ones are, "The love of money is the root of all evil," and "It's easier for a camel to pass through the eye of a needle than for a rich man to enter the kingdom of heaven." When Jesus finds money changers in the temple, he throws over their tables, screaming that the money changers have turned the temple into a "den of robbers."

With the exception of sex, money is the temptation most often cited as leading people to sin. This wasn't just the opinion of the ancient world. If you go onto

Twitter[11] today and post something like, "Money is great," you'll unleash a firestorm of people telling you no, money is at the root of every problem in the world. It's why our environment is stressed, it's why there's so much inequality, it's the reason our politics are so corrupted. We'd all be so much healthier and happier if only we could live in a world without money.

As is the case with many popular, widespread beliefs, the idea that money is the source of all the world's troubles is simply wrong.

Money is arguably the most consequential and transformative human invention of all time.[12] Money shows up early in human history, nearly 5000 years ago. Money is probably the reason we even have "human history." Money's older than writing, and it's very likely the reason we came up with the idea that we could use a string of abstracted symbols to store information. The earliest writing was probably a way to keep track of trades of grain and cattle.[13]

These trades are what makes civilization possible. It facilitates the division of labor. Some people worked with cows, others worked with barley. Spinning off from food production, some people found their time was best spent producing pots or sandals. Even in the earliest societies, specialization of labor drove tremendous increases in the wealth and quality of life

---

[11] Admit it, you don't call it "X" either.
[12] Fans of "the wheel" or "fire" might dispute this, but they can write their own books.
[13] And asses.

for the average person. If you wanted to live in a brick house, wear sandals, drink wine, or enjoy some music, you didn't need to know how to make bricks, sandals, wine, or music. You could take the thing you were good at making—clay pots, for example—and trade this good for another good you needed, like wine.

Of course, this trade economy has limits. It could be that the winemaker has all the pots she can possibly use, and what she really needs now are some roofing tiles. You could go to the person who specializes in making tiles and see if they'll trade for your pots, but that's inefficient. You might have to trade for dozens of things you don't want in order to finally get your wine, and then, after you drink that wine, you need to start the whole process over again.

This is why money is brilliant. You get all the benefits of trade, but you no longer go through the hassle of trading the commodity you specialize in making for the hundred different things other people make. Instead, people who want pots come to you and offer you this new invention, money. Money might be made from something precious and rare, like gold or silver. However, it's not the gold or silver that gives money its value. It's trust. People collectively trust that these little circles of metal or stamped clay are actually worth something.

Once a society accepts that this arbitrary thing possesses a set value, money can be traded for anything. You don't need to worry about whether the winemaker wants your pots. All you need is for them to want your money. And everyone wants your money.

In a civilized world, money is an almost elemental power. It decides the fates of nations and shapes the lives of individuals. It's inextricably tied to freedom and security. If you have plenty of money, you're buffered from many of life's difficulties. It's probably a very rare occasion when you worry about where your next meal is coming from.

Thus, when the Third Horsemen shows up, he's not trying to scare the reader by telling them that there will be drought, or a blight, or that the crops are going to fail. Instead, the disaster is that money is going to be mostly worthless, and a day's wages aren't going to buy enough grain to feed a family.

The not-so-secret weakness of money is that, on its own, it's actually not particularly useful for anything. You can't eat it, build a house from it, or turn it into a comfortable sandal. Money only works because people trust in the authority that issues the money. Some of the earliest things put into writing were contracts. For contracts to work, there needs to be an authority, like a king or a judge, that can reconcile disputes in a manner that feels fair to everyone involved.

Unfortunately, kings and judges throughout history have been known to veer into unfairness, cruelty, and greed. Which is why most civilizations have a separate class of authorities, the priests, who are charged with establishing morals and standards that even kings must follow. If the king follows the morality set forth by the priests, and citizens are behaving themselves, the whole system hums along in relative stability. Life shifts from a day-by-day struggle to find food, and people gain the power to work toward long-term goals. People can build wealth knowing it will still have value when they are old, allowing them to pass it on to their children and grandchildren.

Alas, not everyone will behave themselves. Rather than making pots to sell for money, our ancient potmaker might decide it's easier to murder the winemaker, and grab not only all the wine he desires, but also all her money. The very thing that makes money great at facilitating trade also makes it convenient to steal. It's not particularly hard to carry or to hide, and people desire it enough that the less ethical members of a society won't care where your money comes from.

Rule breakers of this nature, people willing to murder and steal to get what they want, cast doubt on the very underpinnings of social order. It's natural to see why such activities would be classified as being not merely illegal, but also offensive to the gods of a society. Mortal justice can be a fickle thing. Criminals can escape punishment through luck, skill, or bribery. Adding on a higher level of punishment, an afterlife where you will face infallible judgment for your transgressions, can keep enough people in line to keep the social fabric from fraying.

Of course, there are transgressions that don't rise to the level of outright criminality. Infidelity, promiscuity, disrespect toward elders, even simple fibbing. Cheating in one aspect of life can lead to cheating in others, and those who play by society's rules can feel like suckers when they see others taking shortcuts.

And if it becomes apparent that the priests and lawmakers have also been corrupted? Society frays. Once trust in the system is broken, some otherwise manageable crisis, like a drought or an invasion, can cause an entire civilization to crumble.

John of Patmos warned that the world was going to end because sinners had driven the world past the point of redemption. But is that only a warning for the ancient world? Or is it a crisis currently unfolding in our own time?

It's not uncommon for people on every part of the political spectrum to feel as if their political enemies are breaking the social contract. There are also those who feel that the old social contract was a flawed document drafted by the greedy and powerful, and it's time to tear it asunder.

Today, some people resent immigrants crossing national borders without papers. Others look at industries where CEOs grow wealthy while workers strain to pay rent. Our priests and our politicians seem to have given up on promoting morality, and appear content to enrich themselves until the inevitable day that it all falls apart.

Perhaps, in a not-so-distant future, you'll discover your credit cards don't work. The electric grid will collapse little by little, and the internet will come and go. Then, both will simply be gone. There will be no food in the grocery stores, no gas in pumps, and the only lights in the night skies will be the burning of cities.

As all this unfolds, there will be a few geezers sitting on their porch with shotguns feeling immense satisfaction that they saw all this coming. Sinners can enjoy themselves for a long time, but sooner or later, God will have his judgment.

Should you panic? Will this kill you?

It's certainly possible. The dark reality is, history is full of societies where social order broke down. Results are often grim, and death can come on massive scales. In the 20th century, after mankind had collectively figured out how to build tractors, irrigate fields, fertilize the soil, and crossbreed plants to improve yields, 70 million people died of starvation. Usually, this had nothing to do with bad weather and everything to do with terrible politics.

On the other hand, no matter how much old men sitting on their porch might be warning that things are about to fall apart, there's plenty of evidence that the social fabric is rather robust. If famine is a consequence of societal breakdown, what are we to make of the fact that the modern world produces so much food that more people die from obesity related diseases than starvation?

And food production isn't the only achievement of the modern world. Murder and other violent crimes claim far fewer lives than they did during the previous century. For all the fear that our society is growing increasingly hedonistic, statistics show opposite trends. Young people today put off having sex until later ages than their parents' or grandparents' generations. Teen pregnancy rates have plummeted.

Still, even if a frayed social fabric doesn't lead to complete collapse, small failures in the overall system mean you might get murdered, get killed by a drunk driver, or kill yourself with an accidental overdose of an illegal drug.

Will this decimate us?

Maybe. On every continent but Antarctica, we find the ruins of vanished civilizations. The modern world survived the tumultuous 20th century, and emerged as a more peaceful, abundant, and fair world than it was during the 19th century. That said, the 20th century was particularly vicious. There were two World Wars, horrible famines, and one pollution crisis after another.

Despite the countless tragedies of the 20th century, we closed out the final decade with vastly more people alive than we had in the first decade.[14] Technological advancement seems to be keeping the Apocalypse at bay. On a local scale, yes, there are places where a tenth of the population might get killed off by some social breakdown leading to war and starvation, but on a global scale, we're keeping the horsemen at bay.

---

[14] Whether or not this is good news is something we'll tackle in a later chapter.

Could this send us back to the Stone Age?

Unlikely. I said that the ruins of lost civilizations are found on every continent. But you know who's finding them? The residents of our current civilization. The reality is, since money and writing emerged roughly six-thousand years ago, individual societies might collapse, but civilization as a whole staggers onward.

Will this lead to extinction?

No. The primary argument against civilizational collapse leading to extinction is that humans managed to occupy the planet for roughly 200,000 years as hunter/gatherers, small scale agrarian societies, or as nomadic tribes. If civilization were to utterly collapse tomorrow, the survivors might be miserable, but millions, perhaps billions, would find a way to carry on.

Is there hope?

Yes. The British were advised during WWII, "Keep calm and carry on." This is good advice for the current moment.

For all the social disputes that seem to be unfolding in the world, odds are good that anyone reading this book has the power to walk into a grocery store and walk out with several weeks' worth of food without any haggling or hassle. We've fine-tuned the idea of money to such a degree we don't even need to exchange bills and coins. A majority of transactions are now simply an exchange of information stored on computers that might be thousands of miles away from the point of purchase. And... it works! Routinely, seamlessly, and with surprisingly little effort.

If you want to keep famine at bay, the simplest thing to do is to just keep shopping.

F*or the great day of his wrath is come;*
and who shall be able to stand?
Revelation *6:17*

THE SECOND HORSEMAN: COMETS!

Revelation launched us into our first doom, and now we'll draw upon it for the second. As the dooms pile up in Revelation, we reach a passage that describes a plausible threat of extinction. We know it's plausible because this doom has wiped out numerous species before, and will almost certainly do so again. Revelation 8:8 reports, after the second angel blows his trumpet, "…something like a great mountain burning with fire was thrown into the sea, and a third of the sea became blood."

To be written in an age before telescopes, this is a surprisingly good description of what happens when our planet collides with a large asteroid or comet[15].

While there are a few holdouts against the theory that an asteroid impact drove the dinosaurs to extinction, the evidence of a collision occurring right at the end of the age of dinosaurs is pretty solid. Like, literally solid. This is evidence you can dig up, hold in your hands, and examine under microscopes. Not that a microscope is required. It's evidence visible to the naked eye.

---

[15] To save myself from typing "asteroids and comets" a million times in this chapter, I'll be using the terms more or less interchangeably. I apologize in advance to all the astronomers reading this.

In 1980, physicist Luis Alvarez, and his son, geologist Walter Alvarez, were able to show that a thin line of a mineral known as iridium could be found world-wide in the rock layers that mark the final years of the dinosaurs. This division in the rocks is known as the K-T boundary. Iridium is relatively rare on Earth, at least on the planet's surface. It's one of the densest naturally occurring metals, and that density means that most of it sank into the Earth's core long ago, when the Earth was much hotter and more molten.

In meteors, iridium occurs at a much higher ratio than is found in the earth's crust. Thus, the most likely candidate for depositing iridium in a single band that spans the entire world was the impact of a meteor. To spread the mineral so widely, the impact must have been massive. The meteor would have been almost six miles in diameter. A firestorm must have swept the globe when it hit, followed by an era of terrible cold and darkness as ash blotted out the sun. The crater gouged out would have been over a hundred miles across.

You'd think that it would be hard to hide a crater of this size, but, at first, there wasn't a good candidate crater matching the theory. The world has changed a lot in 66 million years. Continents have drifted. It's such a distant age that there were no Himalayas. The tallest mountains of today's world were still rising, and the tallest mountains of that distant era have eroded into little more than hills. You've possibly experienced remnants of these ancient, vanished mountains, if you've walked on a beach. Time has ground them down into the sand beneath your feet.

The history of the world is written into rocks, and geologists have been trying to decipher the messages of these rocks for a few centuries now. So, it's no surprise that, before the Alvarezes started the search for the doomsday crater, someone else had already found it. The structure had been mapped out a few years earlier by geophysicists Glen Penfield and Antonio Camargo. They weren't looking for impact craters. They were working for the Mexican state-owned oil company, Pemex, and were hunting for possible new oil fields. Using gravity maps and magnetic anomalies, they'd traced out a "bullseye" in the Gulf of Mexico. Today, this bullseye is known as the Chicxulub crater, and it's the second largest impact crater ever discovered.

Wait! The *second* largest? Yep. Chicxulub is just one of a fairly long list of craters we've identified that have gouged holes in our planet that are more than 20 miles across. No nuclear weapon we've ever built can rival such destructive power. If the earth collides with another space rock the size of the Chicxulub meteor, it's likely to be the end of civilization, and the odds that any humans survive are rather low.

Though... not impossible. The fact that I'm alive to write this, and you're alive to read this, is proof that even the terrible impact of a Chicxulub class object doesn't sterilize the planet. Alligators, fish, turtles, innumerable plants, and countless insects all shrugged off the impact and survive as modern species only slightly altered from the forms of their distant ancestors. Small, shrew-like early mammals were among the survivors, and they branched off, multiplied, mutated, and eventually filled the world with dogs and dolphins, beavers and bears, lions and lemurs, mice and men. Even some of the dinosaurs seem to have pulled through. You might have one chirping in the branches of a tree outside your window.

The sky is absolutely teeming with rocks, and thousands of them crash into earth each year. The good news is, most of these asteroids are relatively small, and either burn up completely in the atmosphere, or make it to earth as little more than pebbles.

The bad news is, we're certain there are chunks of rock and ice bigger than the Chicxulub meteor, mostly undiscovered. An unknown number of these will cross Earth's path. Another big collision is inevitable. Maybe it will be fifty million years from now. Maybe it will be next year.

Should you panic? Will this kill me?

Don't panic. The odds of a planet cracking asteroid hitting the earth before you finish reading this book are relatively low. The main thing that protects us is that space is unfathomably huge. The ratio of nothing to something is absurdly high, and the dangerous objects of our solar system are more likely to keep passing through nothing than they are to collide with something.

Improving our odds is the fact that our solar system

is also incomprehensibly ancient. It was once far more cluttered than it is today, but over time planets clear out smaller objects in their path, either by crashing into them, or by flinging them far away with gravity.

The rocks most likely to collide with Earth have, over the course of several billion years, already hit us. In fact, the Earth and moon are the product of such collisions. The soil you walk on has its origins in this space debris. So does the water you drink, and the air you breathe. More importantly, *you* are made out of these space rocks, albeit after they've been ground down, baked, cooled, compressed, dispersed, mixed, mangled, and molded into rather complex molecules.

Could this decimate us?

Sure. The impacts of large meteors might be rare, but the risk is still greater than zero. Eventually, we're going to run into another massive rock. If the impact gouged out a hundred-mile-wide crater, it won't be the people living near the crash zone that suffer.[16]

First of all, the shock will ring the planet like a bell. Earthquakes will topple buildings world-wide. Those that don't fall from the quakes might get pushed over by hurricane winds of superheated air. This wind is only phase one of the firestorm. The debris thrown into space by the impact will fall back to Earth, and all of this burning, molten rock will turn the world into an oven.

---

[16] The people near the crash zone will be ash.

Assuming you somehow survive the quakes, the winds, and the burning heat, the ash of the world-wide conflagration means the flames will give way to darkness and cold, as a multi-year winter settles over the land.

If we're only decimated, we'll be lucky.

Can this send us back to the Stone Age?

Definitely. Civilization is, at its core, a machine for producing and distributing food. Farming is going to be rough during the prolonged winter. The infrastructure of our transportation and communications will be toppled and swept away in the first minutes after the impact. Global trade will grind to a halt.

On the off-chance people do survive, they'll be reemerging in a world with no economy, no infrastructure, and with most of the collective knowledge of mankind forever lost. All libraries will have toppled and burned. All the digital archives will be frozen onto slivers of silicon that will be utterly useless. [17]

Can this cause our extinction?

Certainly! The mass extinction of the dinosaurs isn't the only global extinction event likely caused by asteroids. While scientists continue to debate whether

---

[17] If anyone would like to copy all of Wikipedia onto clay tablets, you might be hailed as a great hero of mankind in the distant future.

specific craters are linked to specific mass extinctions, we do have other big impact craters, like the one beneath the Chesapeake Bay in Virginia, or the Popigai Crater in Siberia, that are dated relatively close to the eras of other massive die-offs.

Pinning down specific events, their causes, and their effects, tens of millions of years after they happen is tricky. We may never establish the link between mass extinctions and meteor impacts beyond a reasonable doubt, but the evidence certainly hints at the link, and we only need common sense to understand that an event that can blast a hundred-mile-wide crater into the surface of the Earth is certainly not a good thing.[18]

Is there hope?

Yes? If a doomsday comet appears in the sky tomorrow, we might be in trouble. But, a century from now?

Currently, there are thousands of citizen astronomers all over the planet who go out on cold nights to study the sky. They're hunting for comets and asteroids, and creating an ever-growing catalog of discoveries. NASA and other agencies are also watching the sky, mapping out near-Earth-orbit objects and calculating their future paths. So far, we've found no planet killers.

Unfortunately, just because we haven't found one doesn't mean that one isn't sneaking up on us. In

---

[18] Unless it arrives in an election year. In that case, many of us might welcome a merciful comet of doom.

modern times, we've had large asteroids pass near the Earth that weren't spotted until they were practically close enough to touch.

It's not surprising that we haven't yet found all the potential doomsday objects whizzing through our solar system. Telescopes and the equations we rely on to calculate orbits are less than 500 years old. In the grand scope of human history, this is a rather young field of study.

As we ramp up our search efforts, we're in a fairly small window of time where mankind is aware of the danger of asteroid impacts without having the ability to do anything about them. This window is closing. More telescopes, built with ever more refined technology, will be deployed in the coming years. Searching through the massive reams of information gathered by these telescopes is a task that AI should be able to perform with great accuracy and speed.[19] Assuming we survive and keep refining our technology, in another century it's likely we'll identify every asteroid or comet that poses a danger.

Once we find them, can we stop them? Today, probably not. If we found a planet-wrecker that was going to hit us sometime next year, it's unlikely we could deploy the technology we'd need to deal with it, even though, arguably, most of this technology already exists.

Intercepting an asteroid isn't a matter of aiming a rocket, pressing a button, and waiting for the rocket to

---

[19] Unless the AI rebels and murders us all. We'll examine the odds of that happening in a later chapter.

reach its target a few days later. NASA successfully landed a spacecraft on an asteroid in 2020, but the flight to intercept its target took four years, and the planning took far longer than that. Any asteroid so close that we could reach it in a matter of months would likely be too close to deflect.

For now, the main thing protecting us from a civilization ending impact are good odds and good luck.

Fortunately, as we map out the most dangerous space rocks, we'll probably have decades or centuries to deal with most of them. Then, the same technology that let NASA land on an asteroid a few years ago can carry cargo to the bothersome space rock. We don't need a bomb to get rid of it. Instead, all we need is a small engine. We only need to give the rock a tiny nudge. A distant doomsday asteroid that's sped up or slowed down by a fraction of a second will shift to a new orbit that will miss Earth by millions of miles.

Once again, our best hope of survival is to continue doing what we're already doing. Expand our knowledge, refine our technology, and maintain political systems stable and rational[20] enough to accomplish a goal that everyone agrees is a good thing.

---

[20] Yeah, I know. Rational politics? We're so doomed.

"*The power of population is so superior to the power of the earth to produce subsistence for man, that premature death must in some shape or other visit the human race. The vices of mankind are active and able ministers of depopulation.*"

Thomas Malthus *An Essay on the Principle of Population*

THE THIRD HORSEMAN: OVERPOPULATION!

In our modern world, even the poor are rich when compared to the average person who lived two-thousand years ago. In fact, forget the average person. The poorest resident of a mobile home park[21] is living a life of comfort and abundance that no king could have purchased 2000 years ago. Not that we need to go back two millennia to make this comparison. Let's go back 250 years, and compare the living condition of a poor man today to the life of a king then.

A king of the 18th century might have lived in a palace full of gilded furniture, beautiful sculptures, and a well-stocked library. His living space was vast, with dozens, even hundreds of rooms, and he probably owned multiple residences. Everywhere he went, he was accompanied by servants who cooked his meals, washed and tailored his clothes, and carried his correspondence back and forth to distant kingdoms.

Of course, the letters might take weeks to reach their destination. Months could pass before there was a response. Today, the odds are pretty good that, even if you live below the poverty line, you either own a phone, or can access one. For a small fee, you can instantly talk

---

[21] I mean no offense to anyone living in a mobile home. I've lived in them myself.

with people on the other side of the planet.

Maybe the modern marvel of cross continental communication doesn't impress you. If given the choice between living in a palace a few centuries ago, or a mobile home today, it's not a difficult choice, is it?

It shouldn't be. The mobile home is objectively better. For starters, it has electricity. It's heated in the winter and cooled in the summer. It has a refrigerator, an oven, and probably a washing machine. It almost certainly has a flushable toilet, and running water. Light is available at the flip of a switch. This light is reliable, bright, and clean, emitting no smoke or soot.[22] It's far safer than the candles, lamps, and fireplaces that a king of yore relied on in order to enjoy his library.

And what a tiny library it was! Perhaps you're envisioning rows of shelves stacked high with books. The king might have had ready access to a thousand books. Maybe ten thousand!

This is trivial compared to the libraries available to anyone with a smartphone. There are tens of millions of books available to freely download.[23] The books are better organized, more searchable, and the texts can be scaled to whatever size is needed to be comfortably read. Or, your phone can simply read the book out loud. It can even translate it from a foreign tongue.

---

[22] More accurately, a lightbulb emits no smoke at the location it's used. When we get to the chapter on global warming, we'll explore the dirty emissions of clean light.

[23] Okay, they aren't all free. If you paid good money for a copy of this book, you should take pride in being a person of high morals and peerless taste.

If you're unable to afford a smartphone, you'll have to satisfy yourself by walking into any of the 17,000 public libraries in America. Knowledge has never been so abundantly distributed, and personal wealth is no longer a requirement for access to the writings of the greatest minds of all time.

This is a long-winded way of saying that, as a species, we're doing great. Collectively, we've never been smarter, healthier, or wealthier.

Where does this wealth come from? Human ingenuity is the easy answer. But, ingenuity alone can't produce a lightbulb, a mobile home, or a library. We live in a material world, and all the books and screens and buildings that surround us aren't summoned from the ether. Trace any component of our modern world back to its physical source and you're likely to find a hole in the ground.

Our society is built on the extraction of minerals. Without these minerals, we wouldn't have our mobile homes, let alone our skyscrapers, nor would we have cars, or paved roads to drive upon. The ground is where we also find the fuels we burn to keep our light bulbs burning. Even more important than fuel, the most essential things we extract from the earth are our food and water. We can't eat dirt directly. We need to run it through plants first. Then, many of us wait for the extra step of having animals eat those plants, so that we can eat the animals.

We've gotten really good at extracting food from soil, in part, because of holes in the ground, in this case the holes we drill to collect natural gas. Through some basic chemistry, we can combine this gas with nitrogen from the atmosphere to create fertilizer. This means we're not only extracting resources from the ground... we're mining the air as well.

A scientist named Fritz Haber[24] first perfected the technique of producing fertilizer at commercial scale in 1914. At the time, the population of the world was roughly 2 billion. By 2014, the population had reached 7 billion. As of 2024, we've rolled past 8 billion. In a few decades, our planet will be home to 10 billion people.

At first glance, this is great news. Our population has grown rapidly, but not because women are having more babies. Just the opposite. The worldwide birthrate today is almost half of what it was in 1914. Women no longer have as many babies because we've gotten much, much better at keeping babies (and mothers) alive. We've beaten back the horsemen of famine and disease and created a world where most children survive into adulthood.

But what if this good news is paving the way to a bad future? Every one of the ten billion people who will soon be sharing the earth are going to be consuming resources extracted from the soil, the sea, and the sky. Even with aggressive recycling, there must be some limit to how much copper or iron we can extract. Even with chemical fertilizer, is it possible we'll simply run

---

[24] Haber was acclaimed for making "bread from air." He was also known as the "father of chemical warfare." He was both a benefactor of mankind, and one of its worst villains.

out of suitable land to farm? The seas are large, but big fish grow slowly. We're emptying fisheries with trawlers faster than some species can reproduce. All these fishing vessels, like the trucks on our road, or even the lightbulbs in our home, are mostly powered by fossil fuels. The coal, oil, and gas that has powered modern civilization are the products of millions of years of geological processes. As our population grows, the burn rate of these fossil fuels keeps rising. Warnings of "peak oil" have proved flawed in the past, but mainly because we're getting better at extracting oil from more marginal sources. As our population rises, and as the most impoverished nations grow richer, we'll burn through our reserves of fossil fuels at an ever-quickening pace.

The realization that populations can grow faster than resources can be expanded isn't some new idea. Thomas Robert Malthus foresaw the problem all the way back in 1798, and spelled out the danger in his book, *An Essay on the Principle of Population*. The book laid out a seemingly inescapable problem. Improvements in farming meant more food was available, but more food meant that there would be more people. Since more people would eat more food, all the improvements in food production would literally be nibbled away, and we'd inevitably wind up with more poor, malnourished, and miserable people than we'd had before we implemented the improvements in farming.

Malthus wasn't a voice crying in the wilderness, unheard and unheeded. His theory proved enduring.

By the 20th century the common wisdom among highly educated people was that we were on the brink of disaster due to people's enthusiastic participation in the act of creating more people.

In a 1968 book, *The Population Bomb,* co-authored by Paul and Anne Ehrlich, the doomsday we faced due to overpopulation was explained with stark directness: "The battle to feed all of humanity is over. In the 1970s hundreds of millions of people will starve to death in spite of any crash programs embarked upon now. At this late date nothing can prevent a substantial increase in the world death rate..."

If things were this dire in the 70s, you can imagine how terrible things must be in our current century, when "the population bomb" has gone off, and we've doubled our numbers from 4 billion in 1970 to 8 billion today.

In fact, you have no choice but to imagine it, because the predicted doom failed to arrive. Today, we produce more food per person than ever before. And it's not just food that's more plentiful. Any time some vital mineral grows scarce, we've found ways to recycle it, harvest more of it, or replace it with a more abundant substitute.

Still, we might be making more stuff, but since we're dividing it up among so many people, poverty must be growing, right? Nope. Doubling the population didn't double poverty. In 1970, the number of people living in extreme poverty was 1.7 billion. Today, that number is closer to 700 million. We doubled our population, and still lifted an additional billion people out of extreme poverty.

Where did Malthus and the Ehrlichs go wrong? Simple. They were focused on the known limits of the material world. Oil, iron, cropland, fresh water... the world had finite supplies of all these things. We'd need to strip mine the planet and drain every river to keep up with growing demand. Except, the doomsayers overlooked another important resource, one that has been growing steadily more abundant.

That resource is knowledge.

Malthus could have seen the flaw in his reasoning if he'd been paying a little closer attention. He lived in the era we today call the Enlightenment.

Before the Enlightenment, nearly every difficulty humanity faced was understood as a spiritual problem. If there wasn't enough food, or if diseases were running rampant through the countryside, these happened because God allowed them to happen. If people were suffering, plainly they had fallen out of God's favor.

For thousands of years, it was believed that the suffering we endured in the material world was part of a divine plan. The only remedy to our woes was prayer.

But, before Malthus was even born, some people had stopped looking toward the spiritual world for answers to life's challenges, and instead fixed their gaze upon the material world. They refined their gaze with wonderful instruments, like telescopes and microscopes. They measured, weighed, and tested the properties of matter. Importantly, they wrote down and shared what they were discovering.

I'd argue that the most transformative years known to mankind fell between 1859, when Darwin published his theory of natural selection and 1871, when elements were organized into the periodic table we're still using today. There were centuries of scholarship and numerous precursors to both these revolutionary ideas, but this short window of history decisively shifted mankind's hunt for solutions from the spiritual world to the material realm.

The astounding beauty of matter is that it's both comprehensible and obedient. No priests or mystics are required for understanding or mastery. In science, humankind has invented a system of truth seeking that's freely available to anyone who wishes to investigate the natural world.

With science, ideas that are true eventually win out over ideas that are false. The idea that diseases were doled out by vengeful spirits or dangerous miasmas gave way to an understanding of viruses, bacteria, and the genetic mechanisms of disease.

The reason we didn't starve to death when the world's population doubled is that people had a deep understanding of how plants grow. We understood the genetics of our most important crops, and were able to optimize them to produce more food on less land. We understood the soil, and the chemistry required for it to be productive, and were able to produce fertilizer on an industrial scale.

Another payoff of our mastery of matter was our ability to make and deploy machines that multiplied the effects of human labor. In the not-so-distant past, far more than half the labor of the world went into growing crops. Today, one man and a tractor can harvest enough grain to fill the aisles of numerous grocery stores. Global systems of food distribution mean that crops can fail in one nation, but the supplies of food from other nations ensures that no one goes hungry.[25]

The world didn't spin into famine and poverty after reaching the once unthinkable population of 8 billion. Does this guarantee we'll survive if that number doubles? You don't need to look far to find natural systems under severe stress. I've mentioned overfishing, but our oceans face dangers worse than fishing lines and nets. We've dumped enormous amounts of pollution into the world without much heed to the potential consequences. One of these pollutants is a gas once assumed to be harmless, CO_2. It's a potential doom deserving an entire chapter of its own, so for now, let's wrap up the overpopulation chapter by asking the most pressing questions:

Should I panic? Will this kill me?

No. The forecasts of widespread misery caused by overpopulation haven't come to pass. The reason isn't a mystery. We've rapidly increased our understanding of

---

[25] "No one" is hyperbole. I have no doubt that someone, somewhere, has died from hunger while I was writing this chapter. Still, by every possible metric, hunger has fallen from the ranks of mankind's most pressing problems.

chemistry, biology, and physics, and are able to design and deploy machines that increase the supply of every human need. People aren't a drain on resources. Human intelligence is, in fact, our most valuable commodity. With more people receiving the food, healthcare, and education needed to thrive, our supply of geniuses can hopefully grow faster than our miseries.

Can this decimate us?

So far, the evidence is a pretty firm no.

Can this send us back to the Stone Age?

No. By most metrics, civilization is thriving. Larger populations drive progress rather than hinder it.

Can overpopulation lead to our extinction?

A common analogy used by the doomsayers is to compare humans to yeast. When yeast get introduced into a hospitable environment, they multiply exponentially, devouring everything they can, and spit out alcohol as waste. Eventually, they poison their environment, and die off. Luckily, we're smarter than yeast.[26] We can foresee problems and address them before they arise.

While our species is likely to thrive, the rest of the natural world might suffer. *Homo sapiens* has a long

---

[26] Some of us.

track record of transforming natural environments into something less wild, and more obedient. Look at the American prairie. We drove once abundant species like the bison to the brink of extinction. To make farmland, we killed off the native grasses and stripped the soil of all its original nutrients. We rely on artificial fertilizers and heavy doses of pesticide to coax corn, wheat, and soybeans from otherwise dead soil.

There are little preserves where small bits of prairie have been restored, housing bison that are closely monitored by zoologists. Around the world, wild places are being cultivated, carefully groomed to look like the natural world, yet tamed. We're becoming caretakers of wildernesses that are as artificial as a theme park

Is there hope?

Certainly! Malthus couldn't have imagined that we'd one day make "bread from air." When we do eventually push up against the limits of our resources, we'll most likely be in possession of new technologies to address our problems. It's unlikely our world will reach a point where it's too crowded. If it does get cramped, fortunately, we have a few extra worlds nearby we might transform into new Earths. Though, long before we leave Earth behind, we'll probably leave being human behind altogether. [27]

---

[27] We'll look at this in the chapter on transfiguration.

Some say the world will end in fire,
Some say in ice.
From what I've tasted of desire
I hold with those who favor fire

Robert Frost, *Fire and Ice*

THE FOURTH HORSEMAN: CLIMATE CHANGE!

The 20th century saw a population explosion that seemed to defy the laws of nature. The quadrupling of population failed to produce widespread famine and poverty.

The main reason we escaped doom lies in something that happened long ago. 450 million years ago, the first forests appeared on earth. At that time, Earth's atmosphere had an abundance of carbon dioxide, possibly as much as 6,000 parts per million. The early forests gobbled up this carbon, not just in trees, but in moss and other vegetation. As it died, this vegetable matter sank into vast bogs devoid of oxygen. Over millions of years, this matter grew increasingly compressed, until it transformed into coal.

Similar processes also led to the formation of oil and natural gas. They are called "fossil" fuels because they are the remnants of once living organisms, but these organisms were mostly built out of carbon extracted from the atmosphere, then purified by geological forces. Today, when we burn fossil fuels, we're really burning the atmosphere of our ancient earth.

Of course, we aren't burning fossil fuels for our own amusement. We burn them because they are superior to the sources of energy we once relied on, like wind, water, animal power, man power, and burning wood.

Fossil fuels store energy in compact form. They make the modern world possible, providing us with electricity, powering our airplanes and automobiles, and providing us with the abundant energy needed to transform raw ores into useful materials. Without the decayed remains of ancient forests, we'd have difficulty turning rocks into steel beams or copper wires. Before coal became part of the process, the main fuel that could reach the temperatures needed for smelting iron was charcoal, which is made from trees. The American steel industry was established in the colonial era because we had abundant forests to turn into charcoal. If we hadn't found a replacement for charcoal, today there might be no steel, and no forests.

In this light, fossil fuels have been a boon to nature. They're the reason we stopped burning trees and whales. In the big picture, we're returning carbon that ancient forests captured back into the atmosphere. Where's the harm in that?

It's time to introduce a little thing called "the greenhouse effect." Carbon dioxide traps infrared radiation from the sun, then reemits it in all directions. Heat that might once have been reflected back into space is now spread throughout the atmosphere, warming it. This is a good thing! Without greenhouse gasses, our world would be a frozen wasteland.

The science of CO_2's ability to absorb and radiate heat is uncontroversial. Commercial greenhouses add carbon dioxide to the air, creating warm spaces where plants can thrive.

By burning fossil fuels, we're measurably increasing the amount of CO_2 in our atmosphere. It follows, logically, that this must also be warming our atmosphere. On the other hand, our atmosphere is a lot more complex than the confined air of a greenhouse. Warmer air produces more clouds, which can cool the atmosphere. And, it's not just plants in a greenhouse that enjoy feasting on carbon dioxide. Forests, crops, and ocean flora gobble up some of our waste carbon. Still, in the pre-industrial era, Earth's CO_2 had fallen to about 280 ppm. Today, we're over 400 ppm. Is there evidence that this has warmed the planet?

Yes. While there are difficulties with trying to establish an average world-wide temperature today, let alone trying to pin down the average from 300 years ago, there are numerous lines of evidence that the modern world is warming. Listing them all would require a separate book, and you can find scientists who will gladly dispute any specific piece of evidence. I think it's best not to get bogged down in defending every possible data point. Trust the underlying physics. CO_2 is a known greenhouse gas. We're indisputably increasing the amount of CO_2 in our atmosphere. We can debate how much warming this is causing, and how fast it's occurring, but, long term, increased CO_2 will leave us with a warmer planet.

Is this a bad thing? Maybe. There will likely be several difficulties that will accompany rising temperatures. Sea level rise could swamp some of our most populous cities. Weather patterns could be altered, and we might see historically severe storms and floods, accompanied by sustained droughts in other parts of the world.

Another danger is that warming can create feedback loops to pump even more greenhouse gasses into the atmosphere. If the tundra covering the northernmost reaches of the world begins to thaw, it could dramatically accelerate the warming trends.

The biggest feedback loop of all is us. 8 billion people burn more fuel than 2 billion did. We're adding at least another 2 billion people in the coming years. They'll want cars, crops, internet, and refrigerators. We can, and will, make all of this technology more energy efficient. A modern Toyota makes far better use of the energy in a gallon of gas than a Model T Ford could. This doesn't change the fact that there are more vehicles on the road than ever before, and that number is likely to double worldwide in the coming decades. Houses are better insulated than ever before, but also larger, and filled with more appliances, including energy intensive comforts like hot water heaters and air conditioners.[28]

We use most of our gains in energy efficiency not to cut our energy consumption, but to make it cheaper and easier to purchase even more devices that consume energy.

---

[28] Air conditioning is proof that most people prefer cooler, dryer air, so there's a special irony that the fossil fuels that power them are creating a hotter, more humid world.

The trends are ominous. The science is solid. If we continue to burn fossil fuels... we're doomed.

Or are we?

Should I panic? Will this kill me?

There's a slim chance, but for the most part climate change is unlikely to kill you. You might ask, what about the storms and floods? What about the rising sea levels? By every possible measurement, deaths from weather related disasters have plummeted. Large hurricanes used to routinely have death tolls in the thousands. Today, even the biggest storms that slam into the US don't produce such widespread death. We've gotten better at forecasting and tracking hurricanes. We also have better building codes. A lot of the construction that was most vulnerable to falling in a hurricane has been wiped out by previous storms. New construction is better engineered to withstand high winds and high water.

As for cities vanishing beneath rising seas, vulnerable nations like the Netherlands have been reclaiming land from the sea for centuries. It will require massive engineering projects and a lot of money, but I'm highly doubtful that you'll see a single major city become a new Atlantis.

What about extreme heat? People manage to live and thrive in notoriously hot places like Phoenix and Miami. We have the technology to create pleasant interior temps no matter how hellish it is outside.

The chance you might die from a heat stroke, or a flood, or from a tornado, is definitely a number greater than 0. But, considering all the other ways the world has of killing you, you probably have more immediate dangers to worry about.[29]

Can this decimate us?

No. Humans are adaptable. Pre-industrial humans found a way to live in a staggering array of climates, from frozen northern latitudes to steaming jungles to sunbaked deserts. It's likely that local ecosystems will be altered by changing weather patterns and growing heat. It's just as likely that people can find ways to live with the changed climate, and can actively mitigate the worst effects of the local changes.

Could this send us back to the Stone Age?

Only if that's what we choose. You could make the case that the only hope to avert climate change is to turn our back on an industrialized society. It's appealing to think that, if we stop making use of fossil fuels, we could return to a simpler, more natural way of life. Unfortunately, this wouldn't be a way of life capable of supporting 10 billion people. By some estimates, the human population in the Stone Age was little more than 5 million people. If we simply ran out

---

[29] Nuclear annihilation is still a possibility, as we'll see in the next chapter.

of all fossil fuels tomorrow, we'd enter an era where billions of people would perish from poverty. This is a case where the seemingly obvious solution would be far more terrible than the problem we were trying to avoid.

Is there hope?

Certainly. The best way forward is to keep moving forward. We innovated ourselves into this problem, by tapping fossil fuels to create a massive, rapid, unprecedented increase in human prosperity. We now need to innovate further, and develop technologies that can provide us more energy than fossil fuels with fewer ecological costs.

The biggest escape hatch out of our fossil fuel economy is nuclear fusion.[30] This is the same energy that powers the sun. It produces less pollution than any fuel humans have used to date. The one tiny, itsy-bitsy problem with it is that, so far, despite decades of effort and billions of dollars invested, we've failed to make it work. Mostly. As of late 2024, the number of fusion reactors that have, even briefly, produced more energy than they consumed, is *one*.

Our track record of failing to create fusion power at useful levels isn't the death knell for the technology. In science and engineering, failure is the road that must be traveled to reach success. We might get a chuckle

---

[30] If you're a fan of wind or solar, good for you. I see them as useful solutions for small scale projects in some settings, but am dubious they can be scaled to replace all fossil fuels worldwide.

out of crackling black and white films of crazy early airplane designs crashing before they ever get off the ground, but our modern airlines are possible only because so many people were willing to risk life and limb to test out flawed designs. Today, every fusion reactor design that fails reveals the problems we need to address in the next attempt. Once fusion is perfected, the era of fossil fuels will draw to a close.

Our cars, of course, won't have reactors built into them, but battery technology is rapidly improving. Before the end of the century the last internal combustion vehicle will be retired to a museum.

But what if fusion isn't perfected in the next decade? Or even in the next century? Fortunately, there's a less fussy form of nuclear power, fission. This technology already produces about a tenth of the world's electricity, with zero carbon emissions. Yet, rather than ramping up our use of this power, we're actually shutting down existing plants.

Fission has a bad reputation, and deservedly so. Radioactive waste, the possibility of meltdowns, and the fact that the reactors produce the materials to manufacture atomic weapons are reasons that environmentalists turned against this form of energy.

Before climate change was the media darling of doomsdays, the threat of nuclear holocaust was constantly in the news. It was rational to fear a nuclear war, but this fear likely kept us from adopting a technology that would do less damage to our ecosystem than fossil fuels. The soot from coal plants has easily contaminated more land, water, and air, and caused more human disease and death, than the notorious disaster of the nuclear plant at Chernobyl.

But we don't need to risk another Chernobyl. If you aren't pursuing reactions that are going to make materials for nuclear warheads, it's possible to build reactors from less risky radioactive materials, like thorium. New generations of reactors can also passively cool themselves if something goes wrong. These smaller reactors don't depend on speculative advances in technology, nor are they prohibitively expensive. Unfortunately, due to the still lingering fear of nuclear power, you won't find many US senators eager to deploy them in their home state.

Once more the fate of the planet comes down to politicians making wise choices. While we wait for that day to arrive, step outside. Take a deep breath. It's okay. The atmosphere isn't going to kill you.

Yet.

Back to the struggle, baffled in the strife,
War, war is still the cry, 'War even to the knife!'

Lord Byron, *Childe Harold's Pilgrimage*

THE FIFTH HORSEMAN: WAR!

Sometimes, I go whole weeks without thinking about a nuclear holocaust.

This is a good place to be, mentally, for anyone born between 1950 and 1985. If you grew up during those years, you likely experienced almost daily reminders of the possibility of nuclear war erupting at any moment between the United States and the Soviet Union. Every school I went to as a kid had a bomb shelter. I was 10 when *Planet of the Apes* was first broadcast on network TV. The image of Charlton Heston dropping to his knees in front of the half-buried Statue of Liberty is forever burned into my mind. "You maniacs! You blew it up!"

By the time I entered college, fear of nuclear war was at a fever pitch. I'd talk with friends about what we wanted to do with our futures, and it was surprisingly common to believe that there probably wasn't going to be a future. There were over sixty thousand bombs by then, all much more powerful than the atomic bombs dropped on Japan during WWII. We were told that a single H-bomb could destroy everything in a 10-mile radius, and the fallout would condemn anyone downwind to an agonizing death. Scientists warned that the blast zones and fallout wouldn't be the full extent of our miseries. The smoke and ash of a nuclear war would blot out the sun, creating a nuclear winter where no crops would grow.

If nuclear war broke out, it would be the end of the world, which was kind of the point. The strategy adopted during the Cold War, in the standoff between the United States and the Soviet Union, was officially MAD. Mutually Assured Destruction. There could be no strategic uses for these weapons, no tactical, precision strikes to take out an air base or cripple a supply line. It was understood that if one nuclear warhead was used in battle, it would be followed by a swift exchange of each side's entire arsenal, ensuring that both sides would lose.

Ending the world wasn't science fiction. It was official government policy.

Then, in 1991, the Soviet Union collapsed. In retrospect, it was inevitable. In the first chapter, I spoke about how trust is a building block of an orderly society. But the Soviets system was built on the opposite of trust. They filled their society with secret police. No one could trust their neighbor. Speaking truth became a crime, and lies were rewarded.

If you ran a shoe factory and were ordered to make a certain number of shoes, you'd be punished for telling your superiors that you had no leather for shoes. On the other hand, if you reported that, yes, you'd made all the shoes they'd requested, and already shipped them out, you were praised for your faithful service to the party. In *The Gulag Archipelago* by Aleksandr

Solzhenitsyn, the author documents how badly social order breaks down when people are no longer able to tell the truth about basic facts. The prisons became filled with honest men, and the streets were given over to criminals and imbeciles. Such a system couldn't last forever, so it didn't.

Today, while there are the occasional rumbles of possible nuclear war, the fear of nuclear bombs has shifted away from the MAD strategy of the Cold War to a worry that terrorists or rogue nations might make use of a small arsenal. Everyone agrees this would be terrible, and we might yet see the day when a major city is destroyed in a single blast. But, such a blast won't be immediately followed by a barrage of 60,000 missiles. For one thing, the US and Russia have slashed their arsenals, and it's estimated that there are fewer than five-thousand active warheads left in the world.

That's still enough to create more destruction than we can easily imagine. Submarines currently prowl the seas, with no mission other than to wait for the signal to transform our major cities into bombed out, radioactive hellscapes.

So far, the US and Russia have managed to steer clear of open warfare with one another. But, modern Russia is attempting to restore its former glory by conquering lands it once held in thrall. To boost its military power, it's once more growing its nuclear arsenal.

Will children born today grow up in a world where they, too, are drilled in how to seek cover in their school's bomb shelter? We can only hope that our leaders have the wisdom to turn away from such a path.[31]

Should I panic? Will this kill me?

Having spent much of my life with a dull, aching fear of nuclear war, permitting yourself a few minutes of panic every now and then is a useful strategy for dealing with the stress. Go ahead and get it out of your system.

On the other hand, if you want to go through your life not thinking about this danger at all, that's probably fine as well. Nuclear war is similar to a doomsday comet or the eruption of a supervolcano, which are also disasters that transform the entire world in one rapid shock. But, the doomsday comets and supervolcanoes are inevitable, natural occurrences. There's little doubt of whether they'll happen, only when. Nuclear holocaust requires that someone actively makes a decision that it's time for the world to come to an end. As long as no one makes that decision, we're okay. Probably.

Can it decimate mankind?

It's definitely plausible. We don't even need to release our entire nuclear arsenal. You can imagine a scenario where terrorists detonate a warhead inside a

---

[31] You're probably expecting a snarky footnote to follow this statement, but I'll let you insert your own joke here.

major city. Even if this failed to trigger some doomsday switch launching the rest of the world's arsenal, economic and political chaos would follow. The downing of the Twin Towers in 2001 led to US invasions that resulted in almost half a million deaths, by conservative estimates. A nation seeking to punish the destruction of an entire city could trigger a world-wide conflict that could lead to millions of deaths.

Can this send us back to the Stone Age?

Quite possibly. By now, enough science fiction movies have mentioned EMPs, or electromagnetic pulses, that you are probably at least a little familiar with the concept. EMPs fry circuits and crash power grids. They're useful tools against armies of rampaging robots.

But EMPs aren't the products of science fiction. A nuke exploding at a high altitude creates an EMP that can cripple power grids and take out things like cell networks. Military bases might have sensitive equipment shielded to withstand the pulse, but I'm guessing your refrigerator doesn't. The computer that keeps your Hyundai Elantra running smoothly is also probably not built to military grade. An exchange of EMP attacks might spare us from nuclear firestorms and an accompanying nuclear winter, but still leave us shivering in the dark, hungry, and unable to call anyone for help.

Can this cause human extinction?

Almost certainly not. Even at the peak of humankind's doomsday arsenal, even with the real possibility of a nuclear winter, the primary reason that nuclear war is unlikely to cause human extinction is that there are a lot of people in the world, and we're pretty spread out over six continents and across numerous habitats. If nukes are used, they'll be aimed at important political and economic centers. Rural areas aren't likely to be damaged by the initial strikes.

Even if our industrial society crumbles and nuclear winter does cool large chunks of the planet, it's still hard to see this as the end of the human story. Pre-industrial people thrived in all manner of terrain, and endured all sorts of hardship. Modernity may have softened us, but on a species level we still have the brains and bodies of our hunter-gatherer predecessors. We're fast learners, and, while a post-apocalyptic landscape would no doubt contain undreamt of difficulties, a sizable chunk of humanity will carry on. It might take only a few centuries, maybe even mere decades, before the world once more is at the level of civilization required to start building the next round of doomsday weapons.

Is there hope?

We could go MAD. I don't know whether this should be taken as good news or bad news, but the strategy of

Mutually Assured Destruction seems to actually work. Except for America's use of the first two bombs in Japan, no nation that's developed nuclear weapons has found a way to use them as anything other than deterrents.

What's more, the MAD strategy created a reality where the conventional armed forces of nuclear nations never even risked engaging in conflicts beyond proxy wars. This created a curious paradox. The wars in the second half of the 20th century claimed far fewer lives than the wars fought in the first half of that century, despite growing world populations and despite the fact that we kept building more deadly weapons.

The risk of open conflict between countries with nuclear arsenals required the world to invest real effort into building diplomatic institutions like the United Nations and the World Court. These institutions might be ridiculed as pointless or toothless, but since we've managed to avoid WWIII, maybe they work better than we think.

But, more important than diplomacy, one thing that is making a major conflict unattractive is global commerce. Global trade predates the industrial age, but the economies of the world have never been as interconnected as they are today. The era when it made economic sense to conquer a nation to plunder its resources has mostly come to an end. We've built a world order where information, technology, food, clothing, and all manner of goods flow across multiple state borders under an international system that we've worked long and hard to hammer out.

Your smart phone probably runs software written in the US, on hardware designed in Korea. It might have been assembled in China, incorporating rare minerals mined in Africa. It was possibly shipped to you on a vessel flying a Greek flag, burning diesel from Saudi Arabia. And it's not just high-tech products that draw upon the resources of multiple nations. The shirt on your back might well have traveled through far more nations than you have.

To stave off nuclear war, keep shopping. Keep the world's economies functioning as a single unit. When you order your next Big Mac, eat it with pride, knowing that your rapacious consumerism is helping to keep a nuclear holocaust at bay[32].

---

[32] Unless your rapacious consumerism is destroying the world, which it is. See the next chapter.

THE SIXTH HORSEMAN: TRASH!

We're trashing the planet.

And by "we," I mostly mean "Americans." Which is odd, since at first glance, the United States is almost unnaturally clean. Sure, there's still plenty of litter on the roadsides. In rural areas, I've lost count of how many times I've encountered a creek or a hillside turned into an impromptu dump by the locals.

But, compared to Victorian London, medieval Paris, or ancient Rome, modern American cities sparkle. New York City used to have pigs roaming the streets to devour the waste that was tossed into the roads. Before automobiles took over the roads, the streets of New York were covered several inches deep with horse manure. Also, dead horses! When a horse keeled over while hauling a heavy load, it was often left to rot for days or weeks before being collected. The corpses were sent to rendering plants in Brooklyn, and anything unusable was simply tossed into the nearby bay, a body of water still known as Dead Horse Bay.[33]

Cities have always had a waste problem. Human feces and urine would accumulate on streets and in

---

[33] New Yorkers don't sugar coat things.

ditches. Even after the invention of sewer systems, the sewage would simply wind up flowing into rivers and creeks, fouling the water supply. One reason ancient cities muddled through was that everyone was drunk. Beer making required boiling grains, killing pathogens, and the alcohol and acidity of wine made it inhospitable for microbes.

While the streets of ancient cities would likely strike us as smelly and icky, an ancient Babylonian or Greek produced far less trash than the average person today. More importantly, with the exception of pottery, nearly everything an ancient person tossed onto a trash heap was biodegradable. Food waste, natural fibers, wood, and even early metals were all items nature could recycle. An ancient trash heap would change to fertile soil mixed with shards of broken pottery, given time.

This is no longer true with modern human garbage. The first difference between our waste and ancient waste is that there's simply a lot of it. A typical American produces 1,800 pounds of waste every year. This compares to the global average of under 600 pounds. Our ancient Greeks and Babylonians would have been producing a fraction of this. While the US is an outlier when it comes to trash production, as the world gets wealthier, other nations are starting to catch up.

While the change in quantity is impressive, the change in quality is where things become worrisome. Pre-modern trash was made from things harvested from nature, and nature already had systems for turning organic waste back into soil. Things that couldn't rot, like tin cans or glass bottles, could be reused or recycled. And then…

Plastics. Starting in 1907 with the introduction of Bakelite, plastics began entering our waste stream. Plastics were light, easily molded, and durable. They were also, arguably, good for the environment. I'm currently typing this on a plastic keyboard. The earliest typewriters used keys of ivory. There are billions of keyboards in the world today, and we didn't need to kill a single elephant to make them.

Things once wrapped in paper began to be wrapped in plastic. Plastic was waterproof. It was also sterile when manufactured. It was tough and light. A pack of 100 paper grocery bags weighs about six pounds. A pack of 100 plastic grocery bags weighs only one pound. Given that the US uses over 300 billion grocery bags in a year, the plastic bags can be seen as ecologically friendly simply because we use less fuel in transporting them.

Plus, if you've ever gotten a paper grocery bag wet, you've seen it lose its structural integrity. Run a paper bag through your washing machine, and what will come out will be a wad of mushy scraps. Run a plastic bag through your washing machine, and it comes out looking more of less the same as it did before it went in.

Plastic is unnatural. The reason it can be used to package motor oil, bleach, vinegar, and plain water is that the solvents that degrade other materials don't affect most plastics. The enzymes microbes use to break down other materials have no real effect on plastic.

Which isn't to say that plastic is immortal. UV radiation degrades most plastics, and simple wear and tear can grind them down. The synthetic rubber of your car tires leaves a fine layer of dust behind every time you drive.

This dust is classified as a microplastic. And tires aren't the only source, or even the main source. Anytime you wash clothing made of plastic fibers, some of the fibers wash away with the wastewater. If you've ever seen a building with chipped and crumbling paint, the missing paint still exists, it's just turned into dust and flakes made of, you guessed it, plastic.

In the US, we've gotten very good at hiding our plastic waste. Some of it's burned, most of it's buried in landfills. A big chunk of it, about 950 million tons, gets shipped overseas for "recycling."

Why is recycling in quotes? Because very little of the plastic you toss into recycling bins actually gets recycled. The durability of plastics makes them difficult to break down chemically, and economically every sort of plastic is cheaper to simply make from scratch than to recycle. The plastics that can be successfully recycled need to be isolated from other types of plastic, and separating the different plastics after they've been jumbled together in a single bin is labor intensive. Most of the plastic we ship overseas winds up being buried, incinerated, or ends up lost along the way. Our oceans would probably be cleaner if we stopped shipping plastic waste overseas.

Plastics and oceans are a bad combination. We've already stressed our seas through overfishing, and we're probably adding to that stress by warming them. We also create huge dead zones in some parts of the ocean with polluted runoff. Filling our seas with a slurry of plastic could be our worst assault upon the oceans' ecosystems yet.

It's becoming increasingly common to do a necropsy on a dead whale, or even a sea bird, and find plastic in their stomachs. We know that sea turtles choke on plastic grocery bags, mistaking them for jellyfish.

And as heartbreaking as it is to see animals injured by large pieces of plastic, the more sinister threats are the microplastics I mentioned. We've already passed a threshold where the majority of fish harvested contain at least trace amounts of microplastics in their body. Microplastics get eaten by tiny fish, which get eaten by slightly bigger fish, which get eaten by the larger fish we enjoy having for dinner.

Increasingly, microplastics are part of our diet. How much plastic do you already have inside you? At least a little. Possibly a lot. Just what is it doing to your body? This is still a topic of debate. What's not up for debate is that our bodies didn't evolve to deal with a food chain riddled with micro plastics. If we lived inside a comic book, a diet of industrial waste might give us superpowers. In the real world, we face grimmer consequences.

Should I panic? Will this kill me?

No and maybe. Whatever dangers microplastics pose, they plainly aren't causing people to keel over in the streets. Plastics survive in the environment primarily because they don't get broken down by biological processes. For the most part, our bodies just ignore them. But, there are studies linking microplastics with certain cancers. That said, cancer survival rates are improving. Whatever cancers microplastics cause aren't overwhelming the progress we're making in taming the various forms of the disease. At least, not yet.

Can this decimate us?

Who knows? We might be well adapted to live on land, but Earth is an ocean planet. So far, we haven't traced the extinction of any species to microplastics, but we're still in the early phase of plastic pollution. Soda didn't start getting sold in plastic bottles until the 70s. Plastic grocery bags didn't become widespread until the 80s.[34] The international trade in plastic waste didn't launch until the 90s. The flood of plastic flowing into our oceans is a fairly recent development. What will the consequences be in another century, especially as the rest of the world catches up to the US in plastic usage?

---

[34] I remember a cashier telling me that the store was switching to plastic bags to save the trees.

Can this send us back to the Stone Age?

Possibly. But, it might also drive civilization to new heights. Disposing of waste is one of the primary functions of a civilization. The modern trash industry is surprisingly effective at keeping our trash out of sight and out of mind. Maybe *too* efficient. One reason we think so little about creating waste is that it's swiftly whisked away from our sight to be burned, buried, or barged to another continent.

One can imagine a future Earth where people live inside sealed cities surrounded by acres of greenhouses filled with immaculate gardens, on the shores of long dead oceans filled with sludge. We'll survive by making use of the same technology we'd need to develop to colonize an alien world inhospitable to human life. Only, instead of traveling to another star to find such a planet, we'll simply have made one ourselves.[35]

Can this lead to extinction?

Microplastics have been linked to lowered fertility rates. It's likely that, after a century of increasing population, we're about to enter a century or more of population decreases. Still, I wouldn't place bets against mankind's ability to stagger on into the future. I just worry that future won't be hospitable to many of the species who currently share our planet.

---

[35] With a little grit and elbow crease, humans can accomplish almost anything, by golly!

Is there hope?

Sure. Plastics were created to solve some problems. They made packaging lighter and more durable. They reduce our need to harvest natural materials like wood and ivory. Hospitals are more sanitary with so much of the material coming into contact with dangerous substances like blood being disposable. Our cars and airplanes are lighter, saving fuel.

The challenges of recycling plastic aren't trivial, but they also aren't insolvable. We spent decades making plastics more enduring, but current research is often devoted to making them more biodegradable.

While we're waiting for a technological breakthrough, shop mindfully. Buying items in bulk and storing them in reusable containers isn't going to save the planet, but maybe it will save a single sea turtle, and buy us all a little more time to clean up the mess we've made.

THE SEVENTH HORSEMAN: ALIENS!

When compiling my list of dooms, some popular ideas didn't make the cut. There's no chapter on the "zombie apocalypse," for instance. It's an idea that makes for marketable fiction, but I don't think very many people actually view it as a serious threat.

I was tempted to toss another science fiction premise into the bucket of ideas not to be taken seriously: The alien invasion.

Upon further reflection, I think it's worth a closer look at why so many people do seem to think an alien invasion is not only possible, but likely.

The first question when contemplating an alien invasion is simple. Are there, in fact, alien civilizations far more advanced than our own?

Currently, we have no proof of alien microbes, let alone alien empires. But, we have multiple, well established lines of evidence that gives us reason to suspect that the universe does harbor advanced alien civilizations. Given what we know about biology, evolution, and human societies, it's easy to deduce that some of these civilizations are run by jerks. [36]

---

[36] Most early civilizations were ruled by murderous thieves, though occasionally they were ruled by thieving murderers.

The first solid data point in support of alien life is: We exist.[37] We know that the universe can support life, and intelligent life at that, because we're here to contemplate the question.

A second data point is generally agreed upon by astronomers. There's nothing particularly unusual about our solar system. Earth orbits a yellow dwarf star, which are abundant throughout our galaxy. As stars go, yellow dwarf stars are relatively stable. They stick around a long time. They burn for ten billion years, which is plenty of time for planets to form around them. After a turbulent early era of planetary formation, the planets have at least a few billion years to incubate life and allow for evolution to create complex life forms.

The third data point is a number. A big number. Our galaxy has about 200 billion stars. And our galaxy isn't unique. The universe contains 200 trillion galaxies! That's a septillion stars. That's a 1 followed by 24 zeros. 1,000,000,000,000,000,000,000,000[38]. If only one star in a thousand has the conditions for life, that's a sextillion stars. If civilizations arise on only one in a billion of these stars, there are still a trillion potential space invaders lurking out there!

Which leads us to the Fermi paradox. Enrico Fermi was one of the leading physicists of the 20th century. There's a famous story (perhaps apocryphal) where

---

[37] At least, *I* exist. The rest of you might only be my hallucinations.

[38] A gazillion times a bajillion, to be more precise.

Fermi was having lunch with other physicists. They all agreed that the sheer number of potential homes for extraterrestrial life meant that the universe had to be full of advanced civilizations. Which led to Fermi shouting out, "But where is everybody?"

Fermi's frustration is easy to understand. The numbers tell us life should be abundant throughout the galaxy. Our own world is proof that at least some of this life is capable of evolving into intelligent beings who can build the technology needed to leave their home planets. But actual observation of the universe tells a different story. We haven't found *any* direct evidence of life beyond Earth. All of our neighboring planets show no evidence of intelligent life. They don't even show convincing evidence of simple, single-celled life. When we turn our telescopes to targets beyond our own solar system, we haven't spotted any alien megastructures around stars. Our radio antennas have scanned the skies in vain for a single alien transmission. There's zero evidence that aliens have paid us a visit.

The assertion that no aliens have visited us is certainly disputed. There are numerous people that believe we have evidence that our skies are filled with alien spacecraft. Furthermore, it's claimed that these aliens engage in mischief. Aliens might have helped stone age people build massive monuments. Today, aliens seem to enjoy graffiti, tagging fields of grain with elaborate symmetrical designs. Perhaps their most sinister behavior is kidnapping people, then treating them to vigorous sessions of probing.

Other aliens have visited individuals and revealed ancient wisdom, some of it involving crystals. Does this sound crazy to you? If so, you're in a minority. Pew Research found that 51% of Americans believe that UFOs are aliens visiting Earth. A mere 11% said that UFOs were definitely *not* aliens.

I'm not in that 11%. I think it's unlikely that UFOs are alien spacecraft, and also unlikely that aliens routinely visit the earth. Still, I will gladly admit that, if an intergalactic civilization had the technology to visit, they'd also have the technology to hide from us. My gut tells me it's a one in a billion shot that aliens routinely prowl our sky. But, as we've seen, we live in a universe where a billion is a surprisingly small number.

Still, before I will say, "Yes, aliens are here," I need something stronger than blurry photos, shaky videos, and eye-witness accounts. Eye-witnesses might be solid enough evidence to convict a man of a crime, but science needs something a little more solid. Like, literally solid. There are eyewitness reports of angels, and eyewitness reports of giraffes. Angels remain in the realm of the speculative, while giraffes are accepted as real because, if you make the effort, you can actually go out and touch a giraffe.

The biggest obstacle to aliens visiting us is also the most likely answer to Fermi's question.

Where is everyone? Out there. Like, *way* out there. Far, far away. The universe is so vast that, if it does contain a billion alien civilizations, most of them will

Thirteen Dooms

be a million light years or more from their closest neighbors. No matter how good we get at making telescopes, there's simply no way an optical telescope could ever focus enough light to image planets at that distance. And what if, by sheer luck, we did catch sight of something interesting in orbit around such a distant star? The fastest way of contacting them would be to send a radio transmission, or maybe a laser encoding a signal. Then, it's just a simple matter of waiting two million years to hear back.

The problem is that the universe has a speed limit. Nothing can travel faster than the speed of light. We can propose exotic physics like space warps and wormholes, but the idea that spaceships using such technology can be built is in the realm of pure speculation. The stark, unforgiving fact is that the universe is so vast that it manages to make light somehow seem slow. It's extremely unlikely that spacefaring civilizations are going to frequently encounter one another. The same hopeful numbers that convince us that galaxies are teaming with life are the same terrifying numbers that ensure that we are effectively all alone in the universe. No one is dropping by for a visit.

Given the improbability of alien visitors, why does the possibility of an invasion from space have such a hold on our collective imagination?

The most likely explanation is that we're not actually afraid of aliens. What we fear is ourselves. Humans have been driven to explore new territories to seek more resources since before we were even technically human.

Invasion is in our nature. For most of history, if you found a distant people that had a lot of gold, or tall trees, or fertile farmlands, conquest was an attractive option[39].

Even before history was written down, broken pottery and carved stones and brittle bones long buried in dust tell stories of invasion and exploitation. Once history was written down, a depressing amount of it was devoted to records of war. War soon became a driver of technology. Inventions like wooden shields, bronze blades, bows, chariots, and ships shaped the destiny of nations. The famed conqueror Alexander the Great succeeded, in part, because he armed his troops with spears that were simply longer than the spears every other army was fighting with.

Eventually, the world shifted from bows and spears to the age of guns, and nations that adopted guns had huge advantages over nations that didn't have this technology. The boundaries of modern nations were mostly established during the era when nations capable of making lots of guns colonized and exploited areas of the world that didn't manufacture guns.[40]

We once populated our cities with monuments to conquerors, and their names are still written all over

---

[39] Forget the gold and the trees. Other people were historically regarded as a commodity to be exploited.

[40] This is an absurd oversimplification of a complicated topic, which is kind of my jam.

our maps. But, slowly the monuments are vanishing, and whole nations are being renamed. In an era of greater international connectivity and cooperation, it's slightly embarrassing to be reminded of how much violence, theft, and enslavement underlies our current prosperity.

Some people decry that modern nations now seem to be ashamed of leaders once celebrated as heroes. Behaviors that were once applauded now are spoken of in apologetic tones. Traditionalists might grumble that a nation that's ashamed of its origins is a nation on a path to ruin.

Which is nonsense. One of the many luxuries purchased by modern wealth is an ever-refined sense of morality and justice. Our daintier sensibilities are a luxury good, and while some people might grumble that things have changed for the worse, on vast, measurable scales our new values are making life kinder, gentler, and freer for people all around the globe.

Which brings us to one last, important reason we don't need to worry about alien conquerors. A civilization that reaches the stage where they can send explorers beyond their own solar system is likely a stable civilization that has mastered the art of resolving conflicts without violence. If a civilization can't learn to cooperate and resolve disagreements peacefully, its technology won't be used for space exploration. Instead, the incentive will be to build increasingly powerful weapons, civilization enders like nukes and deadly chemicals and designer microbes.

To throw some more Bible at you, Jesus famously preached, "The meek shall inherit the Earth." A long list of conquerors rather flamboyantly proved this assertion to be wildly incorrect. But, eventually, a little meekness is required to move forward. Only those who are peaceful and cooperative have any hope of spreading beyond their home planet.

Should I panic? Will this kill me?

No. Our greatest protection against alien invasion is reality itself. Space is vast, and the cosmic speed limit is frustratingly slow. A civilization of immortals might contemplate million-year journeys, but, by that stage of technology, if a civilization wanted to spread to new planets, it's probably easier to just build the worlds they desire around the next available star.

Can this decimate us?

See the last answer.

Can this send us back to the Stone Age?

This is a trickier question. Aliens wouldn't need to come here to create upheaval in our world. If, by some chance, we were to discover a civilization on a relatively nearby star, close enough to exchange transmissions and share information over a span of decades or centuries, it would certainly create a great deal of turmoil. A significant portion of the world's population rejects well-established scientific ideas like evolution or the true age and size of the universe.

Would discussions with an alien civilization open human minds to a greater understanding of the universe? Or might some people view the things we learn from the aliens as part of a grand conspiracy, or perhaps even as the deception of a malign supernatural force? In a worst-case scenario, I can imagine a backlash against science and technology.

That said, I imagine that the most likely result of alien contact is that most of us would shrug and continue living the same lives we'd been living before the aliens signaled us.

Can this lead to our extinction?

We probably won't go extinct from an alien invasion. But, it's possible that, if and when we do finally encounter alien life, it might stoke our desire to leave Earth behind, and spread out among the stars. But will we go out among the stars as humans, or as something else? Our current bodies aren't well designed for life on other words. But, as our knowledge of genetics and our ability to manipulate molecules continues to grow, why should we remain stuck as simple *Homo sapiens*? We'll explore this further when we reach the twelfth horseman.

Is there hope?

Given the extremely low odds that any aliens are out there wanting to conquer us, I think we can remain optimistic that this won't be how mankind meets its doom. But, this has been a chapter that tossed out a lot of big numbers. Maybe, the sky is full of trillions of civilizations, and one of these is devoted to making the galaxy great again by getting rid of us before we start shooting trash into orbit. What then? The best tool in our defense tool box is knowledge. The more intelligent we get collectively, the tougher we'll be to conquer.

Unless our quest for intelligence has already doomed us. That's a distinct possibility we'll look at when we reach the ninth horseman. Before we reach him, though, we need to consider our vulnerability to another sort of invasion. And, unlike the aliens, we know these invaders are already present on our world. They're already in your house. In fact, they're swarming over you *right now!*[41]

---

[41] *Dun dun dunnn!*

THE EIGHTH HORSEMAN: PLAGUE!

You might remember there was a bit of a kerfuffle back in the early months of 2020. Videos out of Wuhan, China, showed people suddenly keeling over dead. News soon followed that hospitals were overflowing in Italy, and in a matter of days a new word had entered out vocabulary: *Coronavirus*[42].

Officials on TV told us not to panic. Then, the virus reached America. After a few weeks of telling us to remain calm, the officials switched to a new message: "Okay, now you can panic." They didn't put it quite so bluntly, but they did tell people to stay home, and shut down bars, restaurants, schools, libraries, and even public parks. Liquor stores, by some leap of logic, remained open.

Overnight, there was no toilet paper in stores. If you ordered stuff delivered, you were advised to wear gloves and a mask and wash packages with bleach to make them safe.

There were refrigerated trucks parked outside hospitals to hold the bodies since morgues had run out of space. There was talk of digging a mass grave in Central Park.

---

[42] New to the average person, at least. Scientists had been studying them since 1965.

Day by day, the number of dead were reported, first in the dozens, then the hundreds, then the thousands.

It felt as if the apocalypse had finally arrived.

It hadn't.

True, by the end of 2020, 350,000 Americans had died from COVID-19. This only made it the third leading cause of death in the US that year. Almost twice as many died from heart disease. Cancer was the second most common cause of death.

Every death was tragic, but in the end COVID-19 didn't fill mass graves. It was the worst respiratory disease to have swept the world since 1968, when a strain of flu emerged from China that went on to kill as many as four million people. (Equal to more than eight million today, since the world population has doubled.)

The good news is, COVID-19 wasn't as terrible as predicted. Life has returned to normal for most people, though you still might encounter people wearing masks in public.

The bad news is, if another virus does arise, we might be even less ready for it. All the people who believe we overreacted to COVID-19 will be less inclined to take precautions when the next deadly virus emerges.

The even worse news is, it's almost a certainty we will one day face a virus more deadly than COVID-19. Just how deadly can a virus or other microbe get? The answer is a grim one. Microbes have unleashed mass death again and again. Bubonic plague, also known as the Black Death, broke out in Italy in 1347 and went on to kill a third of the population of Europe. When smallpox was carried to the Americas by European sailors, it's estimated that upwards of 90% of indigenous Americans died from the disease.[43]

Nor do we have to reach back in history to find diseases with devastating death tolls. In recent decades, over 36 million people have died from HIV.

Why is it almost inevitable that deadly microbes are going to cause us problems long into the future? Natural selection. Viruses need hosts to survive. Their hosts have evolved immune systems to keep the microbes at bay. Unfortunately, pathogens mutate, producing variant strains, some of which can elude the immune systems of hosts. With a few lucky breaks,[44] viruses can mutate into something that can infect an entirely different organism. This is how a bird virus can spread to pigs, then from pigs to humans.

Perversely, thanks to natural selection, all the tools we use to battle microbes can, over time, make the microbes even stronger. Antibiotics work well for years, until resistant strains emerge. Vaccines for things like the flu or COVID-19 can never provide permanent protection, since viruses that spread rapidly also mutate rapidly, giving them multiple pathways at thwarting a boosted immune system.

Fortunately, mutation and variation work in both directions. Human immune systems vary due to genetics, environment, and past exposures. Even

---

[43] This is a high estimate. The lowest estimate? 50%.
[44] Lucky for the virus obviously, not the hosts.

dreaded diseases like the black plague and smallpox failed to infect every member of a given population. People who survive infection emerge with immune systems calibrated to provide robust protection against reinfection.

Still, if we wind up encountering a virus that spreads through the modern world with the same fatality level that smallpox afflicted upon the indigenous Americans, it's fair to say the consequences would be apocalyptic. Anything that wiped out 90% of the population in a matter of a few months would cripple economies, leading to even more suffering and death.

Should I panic? Will this kill me?

Panic could potentially save your life. Years after COVID-19 failed to deliver doomsday, there are people I know who still avoid it like the plague.[45] For some, the precaution has paid off, and they've simply never contracted the virus. Microbial disease is one of the rare cases where extreme, even irrational fear, might keep you alive. You don't even need to become a recluse who only ventures outside garbed in a full bio-hazard suit. Even mild precautions like staying home more and steering clear of crowded public spaces can tilt the odds in your favor.

---

[45] It would be a waste not to use this cliché somewhere in this chapter.

Can this decimate us?

No question! COVID-19 death rates didn't hit 1%, let alone 10%, but the Black Death and smallpox are there to remind us what a virus can do if it shows a little grit and gumption[46]. And these viruses emerged in a time of relative isolation, when travel between continents was often a journey of weeks, or months. Today, a person can catch a virus on one continent in the morning, and by the evening they might be spreading it in a nation a thousand miles away.

Can this send us back to the Stone Age?

Maybe. Smallpox probably ended the Aztec's civilization. The disease killed far more people than conquistador's guns. On the other hand, plague doesn't harm computers, machines, or books. Even if some new plague did carry off 90% of humanity, all the tools and information needed to maintain civilization remain intact. And Europe actually experienced an economic boom following the Black Death, as a shortage of workers led to rising wages and a growing reliance on automation and machinery. So, while there are cities hidden beneath jungle vines to remind us of the destructive power of plagues, every factory you drive past can trace part of its lineage back 700 years to

---

[46] Viruses are proud of their work ethic.

when some flea-bitten rats arrived in Europe.

Can this lead to extinction?

Not likely. The same force that creates plagues—mutation—also creates resistance. Viruses and bacteria aren't new threats. Our immune systems evolved to combat these things long before we were human. Our earliest ancestors had immune systems before they developed more advanced structures like eyes or a spine. Natural variability in the human population means that there will always be some people that simply don't fall ill from even the most virulent disease.

Is there hope?

First and foremost, we need to keep studying both viruses and our immune systems. So far, the dream of a universal vaccine that halts all viral disease is beyond our grasp, but there's no reason it must remain so forever. Viruses, while tiny, are still nothing but matter, and we humans are increasingly proficient at manipulating matter. Our current collective brainpower has so far failed to crack the code, but, fortunately, we've arrived at the era where we're rolling out new and improved forms of intelligence.[47]

The other thing we're doing well is traveling and

---

[47] Artificial intelligence is the great hope of mankind. Unless it dooms us. See the next chapter.

intermixing populations. *Hold on!* Doesn't this just help viruses spread faster and further? Yes. And this is a good thing.

Many of the plagues of the past had such devastating impacts because they were introduced into populations that had no immunity. There were European's who could survive smallpox infection well enough to cross an ocean while infected with it. Europeans had long centuries to evolve defenses against the disease, and the disease had just as much time to find new ways of infecting its host. When people who'd never been exposed to any strain of smallpox encountered the disease, it was unimaginably lethal.

Today, populations are less isolated. We're swapping more diseases than ever, but there are few pools of people with absolutely no exposure to a broad range of microbes. We're still at risk from viruses that evolve in animal hosts, but microbes no longer have centuries to circulate among humans on one continent before unleashing misery on new lands.

We might never completely win our battle against pathogens. But, thanks to fundamental biological laws, we'll almost certainly never completely lose, either.

*The darkness drops again; but now I know
That twenty centuries of stony sleep
Were vexed to nightmare by a rocking cradle,
And what rough beast, its hour come round at last,
Slouches towards Bethlehem to be born?*

William Butler Yeats, *The Second Coming*

THE NINTH HORSEMAN: ARTIFICIAL INTELLIGENCE!

Artificial intelligence is a staple of science fiction. It turns against humanity in movies like *The Matrix* and *The Terminator*, or, on a smaller scale, is willing to endanger humans to accomplish a mission, like in *2001*. These malign machine minds are counterbalanced by helpful, even kind, artificial intelligences, like Data from the *Star Trek* franchise or C-3PO and the ubiquitous helper bots scattered throughout the *Star Wars* universe.

Whether friend or foe, I always assumed these thinking machines wouldn't arrive until some far future date.[48] I thought we might see some progress in some tasks. I could imagine self-driving cars, or going into fast food restaurants where there were no humans on the staff.

But a computer I could have a natural sounding conversation with? A digital mind that could produce poetry or music or art? Impossible. Creativity and communication were such innately human traits, it seemed unthinkable machines would invade these domains before they perfected more mundane tasks, like sorting and folding laundry.

---

[48] Like 2001.

Yet, unless you're living under a rock, you're already seeing AI generated content in numerous forums. If you ask a question of Google or Bing, AI is involved in supplying you the answer.[49] The 2024 election was also the first election where we saw digital clones of actual politicians delivering faked sound bites. We've been used to robots replacing blue collar workers, but now AI has computer programmers and even surgeons worried about their futures. In recent tests artificial intelligence AI often outperforms human doctors in diagnosing certain ailments.

My digital butler hasn't appeared yet, but digital assistants on my phone can answer my questions about weather, math, and basic nutrition. Google's AI doesn't always give me accurate answers to simple questions,[50] but it's a champ at mapping driving routes that will produce optimal fuel efficiency.

So far, none of this tech is perfect. I'll ask the Echo in my kitchen some perfectly mundane question—"How many calories are in an avocado?"—and get back utterly random answers—"The Alamo is the site of a battle in San Antonio, Texas."

Thanks, Alexa. Super helpful.

Still, for all their flaws, these mechanical minds are increasingly part of our everyday lives. Computer technology evolves at a rate that makes viruses look

---

[49] Frequently, a laughably wrong answer, and occasionally, a dangerously wrong answer.

[50] For instance, if you ask Google, "What is the greatest threat of human extinction," it answers "AI." Such a kidder.

like slowpokes. With each passing month, AI gets better at mimicking human language. It can write essays with ease, and is increasingly capable of writing poems, screen plays, and even novels. That's bad news for someone like me, who pays his mortgage by writing books. I thought a creative career would be safe from a robot revolution. I was wrong.

Nor are novelists the only people with careers in the crosshairs. AI illustrations are already turning up in advertising and package design. But a world where novelists and illustrators have difficulty making money isn't really a very different world than the one we already live in. For the last few decades, whenever industries have been made obsolete by computers, the refrain from Silicon Valley has been, "Learn to code." But AI can already write code. And it's not only the digital world undergoing disruption. AI can design circuits, improve fuel efficiency in cars, and draw up blueprints to build better robots.

Oh, right, the robots. A digital mind in isolation has its uses, but it needs to take on a physical form before it can really resemble the intelligent machines of science fiction. Even before AI, modern robots were increasingly becoming more agile and adaptable. Humanlike robots can perform feats of balance and flexibility that were once the domain of circus acrobats. Not that most robots will take the form of humans. Our skies and streets are increasingly filled with drones.

It doesn't take much imagination to see where this all ends. If today you fell into a Rip Van Winkle like sleep, and slept for twenty years, you'd wake to a world that is seemingly run by ghosts. The cars on the highways have no one behind the wheel. Lawnmowers crawl across lawns with no one pushing them. The sidewalks are bustling… but not with people. The world is now populated by robots, and they're all staring as if you've escaped from a zoo.[51]

This is, of course, a rather extreme place for imagination to take us. A more immediate worry isn't that robots are going to kill us. It's that they'll be able to do our jobs better than we can.

It's difficult to think of any skilled human trade that's completely safe. Before the end of this century, there likely won't be many human brain surgeons. A robot will be more skilled at diagnosing and pinpointing the portion of your brain requiring intervention. Instead of sawing your skull open, an AI will likely guide robots smaller than gnats through your arteries to perform whatever repairs are needed. These tiny robots might be delivering payloads of drugs directly into the cells that need them. The drugs will have been designed to your individual needs by AIs.

It would almost be a paradise, except, without jobs, how will we ever pay for anything?

Of course, this isn't the first time we've confronted a technological revolution. For centuries, people have been warning that machines were going to take jobs from humans. For the most part, they've been right.

---

[51] *"A wild human! Someone call the human catchers!"*

To distribute this book in the year 1400, hundreds of scribes would have been employed to copy the manuscript. There aren't many people making a living as scribes today. There also aren't many people meticulously placing metal letters in even rows to typeset a page of prose, nor many humans laboriously sewing printed pages into a somewhat ragged binding.

Machines are better at nearly everything. They obviously eliminate jobs that require brute strength. You don't see teams of men dragging large blocks of stones up ramps to build interstates. Repetitive tasks are also perfect for automation. Any small task that needed to be performed hundreds of thousands of times to produce something useful, like weaving cloth, or making sheets of paper, was something ripe to be replaced by a machine.

The remaining bastions of human employment were jobs that required intelligence, creativity, or good social skills. But, cashiers, waitresses, and bank tellers are being replaced by touch screens. Trades that required a lot of education and experience (like law or medicine) seemed beyond the reach of the machines, but the next generation of PhDs might find they have stiff competition from artificial minds with perfect memory and the ability to work tirelessly at lighting speed around the clock, seven days a week.

We've been here before. In 1811, in England, riots broke out around textile factories. New machines were being introduced that could automate processes once done by workers. Word spread that a man named Ned

Ludd[52] had smashed the machines that had taken his job, and this inspired other men to take up hammers and break the machines.

Their plan didn't work, thankfully. We live in an era of unprecedented prosperity, a prosperity that simply couldn't exist without machines.

But it's one thing to lose a job to a mechanical loom. Setting people free of mindless factory labor unleashes fresh waves of human creativity and innovation. But what will we do when machines become better than we are at imagination?

Should I panic? Will this kill me?

It's extremely unlikely that AI will turn against humans and hunt us down. That said, AI and robotics can quite plausibly replace a large majority of current human careers. Every job from manual laborers (like fruit pickers) to service workers (like cashiers) to knowledge workers (like pharmacists) are likely targets. Workers with once valuable skills, like engineers, and creative geniuses, like doomsday book authors, also face uncertain futures. AI doesn't need to be murderous to create societal upheaval.

Can this decimate us?

Maybe? One real danger of the rise of AI and robotics

---

[52] Ludd was likely apocryphal. No historian has ever proven he existed.

is that it opens up a new form of combat. Robots won't decide to get rid of humans. Other humans will decide to get rid of humans, and robots will be their weapons.[53]

The US has carried out thousands of drone strikes since 2001. Casualty numbers are disputed, but it's safe to say that thousands of civilians have been killed in these strikes, perhaps tens of thousands. As I type this, the ongoing invasion of Ukraine by Russia is a war where drones have emerged as a major weapon on both sides, with casualties again climbing into the thousands.

Currently these drones rely on human operators, but autonomous drones capable of picking targets on their own are technologically feasible. It's not difficult to see a MAD scenario developing. The US builds millions of autonomous killer drones so that China and Russia can't use their millions of autonomous killer drones without provoking doomsday. Only, unlike nukes, which required scarce and difficult materials to manufacture, and a huge infrastructure to deploy, killer drones can be 3D printed by a misanthropic nutjob in a suburban garage.

Decimation is unlikely, but nightly casualty reports of the unending drone wars might become a routine part of our future.

---

[53] "Murderbots don't kill people, people kill people!"

Can this send us back to the Stone Age?

Returning to Ukraine, drones are being used to target power plants, water mains, and server farms. It likely won't end civilization on a planetary scale, but robot warfare could definitely be used to create pockets of the world where modern infrastructure crumbles.

Can this lead to extinction?

We can again imagine a scenario where AI decides that humans are a threat that needs to be eliminated. Or, if not a threat, just an old, inefficient technology that no one needs around anymore.

Luckily, humans do have a few advantages that might permit us to survive a robot apocalypse. Our most obvious advantage is that it's ridiculously easy to make new people. This has always been our most powerful weapon against any apocalypse. Plague, war, famine… these great enemies of mankind have never claimed so many lives that we can't simply replace the dead within a generation or two. Humans are self-replicating, and are able to fuel themselves on just about any landscape.

Robots, on the other hand, require modern infrastructure for assembly. You need minerals gathered from all over the globe, as well as copious amounts of energy. Nor are machines very efficient at repairing themselves. A human body can run for many decades with all its original parts, but not many cars can go a full decade without something needing to be replaced. Machines are also finicky about their fuel sources. Try putting diesel into your leaf blower and see how long it runs.

Humans can survive without elaborate global infrastructure. For 200,000 years, our most advanced tools were sharp rocks and pointy sticks. Robots make sense in an orderly, modern world, but humans can switch between civilization and wilderness with surprising ease.

For this reason, if a human/robot war ever did play out, I think the robots could win the battles, but lose the war. Infrastructure is always a target in war, and robots are even more vulnerable to the effects of sabotaged fuel pipelines and bombed power plants than we are. A war wouldn't last very long before the robots fell into disrepair. Humans would start anew among the ruins.

Is there hope?

Preventing the rise of killer robot armies probably comes down to treaties and regulations. The ongoing conflicts of today are driven more by ancient feuds than the exploitation of resources. Trade, advanced technology, and the free flow of ideas and information gains a country more wealth, prestige, and power than invasions can produce. Plainly, not everyone has gotten this message, but the historical trends give us reason for hope that wars will one day be found only in the historical record.

Let's forget the killer robots. Are all human jobs going to be destroyed by AI? Have we created a technology that creates more poverty than prosperity?

I have some actual experience with this topic. As I'm writing this book, I'm also reviewing illustrations from an artist I've worked with on several projects. He's one of four artists I'm paying for art this year. I've developed relationships with artists who produce great work and bring a bit of themselves into each project. I can spend more money on human artists this year because I operate a successful indie publishing business that turns a profit.

Part of that profit is made possible not by the books I publish, but by buttons. I design and sell novelty buttons and magnets, and currently offer over a hundred designs. A lot of these buttons feature only text, some feature commercial clip art, and some make use of my own drawings or photographs. But, more and more, I'm using AI to generate the art I need. It's faster than drawing it myself, and no individual button sells enough that it would be profitable to hire an artist at professional rates.

AI lets me get a good design out quickly and inexpensively. This boosts my bottom line, letting me pay better rates to artists and to the authors I publish in my anthologies. By saving me time and money in creating buttons, AI allows me to spend more time and money on creating books.

But, what if, in the very near future, people are

publishing entire novels written by AI? Don't I fear for my livelihood?

Not even a little bit. The not-so-secret secret of publishing is that writing books isn't some tremendously rare skill that few people can master. The world is full of authors, millions of them, and absolutely flooded with books. Every book I write is already in competition with entire libraries of existing books. If you Google how many different books there are in the world, one of the results is the curiously precise calculation of 134,021,533[54] on the website Mental Floss.

If AIs churn out another 10 million books next year, I don't see how it changes my world at all. Writing books is surprisingly easy. It's mainly just daydreaming, research, and a little typing. The tricky part of making a living as a writer is getting people to actually read your book.[55] This requires learning a whole host of skills separate from actually writing. It's more than just marketing. It requires putting in a lot of hard work connecting with other human beings so that you can learn who will want to read your book, and have some idea why they want to read it. This in turn requires a further journey, into one's own mind and heart, so that you can understand why you want to write a certain book, and know what it is you're hoping to accomplish with your words beyond making

---

[54] Which means you're reading book # 134,021,534.
[55] Thanks for reading this book!

money.[56] An author lazy enough to let a computer write their book for them isn't going to put in the work needed to actually connect with readers.

If you want to keep machines from driving humans out of art, lean into being human. Connect with other creators. Form relationships and partnerships.

When I'm out at shows selling books, before the event I'll wander through the artist alley, hunting for artists whose work catches my eye. The majority of artists I hire are people I met at these shows. Three of my book narrators are people I met in person at events. They came up to my booth, handed me a card, and just flat out asked if I needed narrators. I don't always need their services, but when I do get around to recording a book, I choose someone out there hustling for business over someone sitting back and passively waiting for me to find them.

Machines might master producing products, and even creating new ones. But, at its base, our economy still hums along on human relationships. The thing that guarantees that people will continue to thrive in our future economy is our ability to connect with other people. The keys to success are empathy, sincerity, warmth, and honesty. For now, at least, people can fake these better than any machine.

---

[56] Because, odds are, you won't make much.

THE TENTH HORSEMAN: NATURAL STUPIDITY!

One reason corporations are so diligently working to perfect artificial intelligence is that natural human intelligence seems to be pressing up against some built-in limits. It wasn't that long ago that mankind's highest technological achievement was a well-broken rock. This is no minor feat!

Very few residents of the modern era could be dropped into a forest buck naked and hope to survive. But, our Stone Age brethren did more than survive. They thrived, spreading out of Africa to occupy every continent except Antarctica. We found ways to live in blazing deserts and frozen wastes. We mastered survival in jungles and grasslands. When it came to making the most of the natural world, we were geniuses!

Now, we live in a highly *unnatural* world. We were collectively smart enough to figure out some of the fundamental rules of material reality. We solved the riddle of electricity. We crafted machines to replace human labor, and went on to make machines that replicated elements of human thought. The written word can be thought of as a mechanical system for storing memories. And the cheapest, simplest pocket calculator can add, subtract, divide and multiply better than all but a handful of savants.

With these tools enhancing our natural gifts, we've performed wonders. We speak to people on distant continents using pulses of light. We used controlled explosions to lift things into orbit, then launch them on astonishing journeys to fly by planets at the edge of the solar system, and even to transmit data from the realm where our solar system bleeds into interstellar space. These sort of engineering problems are essentially binary. We get the calculations right and succeed, or we get them wrong and fail. There's not much wiggle room.

Unfortunately, many important human questions don't present such clear-cut answers. We have to muddle along with confusing and conflicting data. I'm a well-read person, with a reasonable level of scientific literacy, but if you asked me to describe something as simple as the perfect human diet, I'd struggle to do so. Throughout this book, I've talked about economics and trade, but if you asked me to invest $10,000 for you so that it turned into $1,000,000 in a few years, I doubt I'd do a very good job of it. On aspects of modern life I deal with every day, economics and diet, I'm frequently baffled. There's so much information available, and it can be difficult to discern fact from opinion, and informed speculation from outright falsehood.

In the words of Yeats, we live in a world where "the best lack all conviction, while the worst are full of passionate intensity." In many areas of life, the loudest voices are often shouting out things that are simply untrue. As a species, we are often seduced by prepacked world views underpinned by seemingly logical arguments. When facts emerge in conflict with these ideas, we often turn our backs on reality, and choose to live in our carefully constructed edifices of sense-making.

Is it possible that, as our technology gets smarter, as a species we're collectively getting dumber? When I was young, pocket calculators were turning up in schools. Math teachers warned that, by relying on calculators, we'd create a world where people could no longer do math in their head. The intervening decades have proven them mostly correct. Innumeracy is widespread.

A further warning came when the internet blossomed, and people started googling answers to common questions. Crowd sourced databases like Wikipedia often contained flawed information, or were vulnerable to manipulation by people with an agenda. I remember hearing as early as the nineties that the internet was going to become a tool where misinformation was more abundant than truthful information. People warned that staid, boring facts would get buried beneath a barrage of sensationalism and lies. A few minutes online reveals that this, too, has come to pass.

We live in a fog of paradox. We have more accurate information at our fingertips than ever before, yet routinely embrace and promote falsehoods and fictions. We exist in a world where more calculations are being performed each second than at any time in history.

Math underpins modern life. Yet, somehow, we constantly encounter individuals who have difficulty performing once mundane, daily calculations, like making change.

And weakening math skills are the least of our worries. Well-established science is often outcompeted by nonsense. Many people don't believe in natural selection. Others think vaccines contain microchips. Prominent politicians spout insane conspiracy theories about a cabal of elite world masters steering hurricanes, and, instead of being laughed at, get voted back into office.

But truthfulness and politics have an ever-weakening relationship. In the mouths of congressmen and the pundits that pontificate on their actions, words routinely get stretched so far from their original meanings that they come to represent the exact opposite concept.

Over centuries, we've managed to build a highly complex world on a foundation of tested science and verified information. Instead of attempting to understand this complexity, a lot of people embrace things that sound right without seeming to care very much whether these things are accurate.

Can we run a complicated world with our supply of increasingly simple people? Or, have we pushed the limits of our monkey brains past their natural limits, leading to our inevitable doom?

Should I panic? Will this kill me?

Stupidity is dangerous. A diner at a restaurant who can't figure out a tip probably isn't going to get anyone injured, but a driver who blindly follows GPS without bothering to actually watch the road might kill someone.

Still, let's keep some perspective. Aristotle was complaining about the stupidity of youth 2500 years ago. The reality is, most of us are as smart as we need to be. Very few of us need to send rockets to other planets. Most of us navigate the paths of daily life without once tripping over a square root. If you're smart enough to use a microwave, you're likely smart enough to survive the modern world.

And you're probably pretty good at using your microwave. You can't explain exactly how it works, but you likely understand what can be easily prepared in it and what won't cook well. I doubt you shove whole chickens into it and hope for the best.

Let's swap you back in time to the era of a great-great-great-great-grandparent. You find yourself in a kitchen with a large fireplace. There are hooks in the fireplace, a variety of iron pots, and off to the side there's a stack of wood and a variety of tools for tending and maintaining the fire. There's no refrigerator. There's not even a can of Spam lurking in a cabinet. Instead, there's an unplucked, headless rooster hanging by its feet just outside the kitchen window.

Since you're hungry, and have no idea how to get back to the modern era, you bravely try to make dinner.

To anyone watching, you probably look like an incompetent dimwit. Assuming you manage to get dinner on the table, it will probably be overcooked, or undercooked, or somehow both at the same time.

Technological advances leave us increasingly ignorant of the skills that were needed to function in previous eras. If you're sixty, you might find yourself standing at a cash register in front of a teenage cashier. You just handed her a $20 bill and a quarter to pay for your $17.23 sub.[57] You roll your eyes as she uses a calculator to figure out what change she owes you. Meanwhile, she's rolling her eyes, wondering why you're wasting her time by handing her cash instead of just tapping your phone against the Square like an ordinary person.

In this light, the common suspicion that everyone is increasingly ignorant is both completely accurate and completely normal. Younger people don't bother to learn once-common skills because they no longer need them. Older people don't comfortably learn new skills, because they've managed to live for decades without them. If all the younger people you meet seem a little dim, and you yourself can't understand half the stuff they're jabbering about, take heart. You aren't witnessing an excess of stupidity. You've actually encountered a bubble of poorly distributed genius.

---

[57] Or maybe a denarius for a handful of wheat

Thirteen Dooms

Can this decimate us?

Unlikely, but maybe? Here, all our other doomsdays blend together. We have fantastic modern tools for fighting plague, but if the public grasp of science is so degraded that people avoid vaccines, or are unable to meaningfully evaluate risks, it's not hard to see how a truly deadly plague could spread even if there was abundant, accurate information on how to avoid it. And by this point in the book, you might have gathered that I worry that some modern politicians might be, to put it politely, somewhat underqualified for their positions. No matter what your political leanings, you've probably noticed there are some real boneheads in elected offices who get to vote on policies about killer drones, nuclear power, and the next deadly virus.

One well documented case of stupid politics leading to mass death is the Great Sparrow Campaign that unfolded in China in the 1950s. The problem? It was believed that sparrows ate grain in fields, lowering crop yields. The solution? Kill the sparrows! The result? Without sparrows eating them, the locust population exploded, leading to widespread crop failure. The consequence? 20 to 50 million people died of starvation. That's a lot of people dead because someone in power didn't understand basic ecology.

Can this send us back to the Stone Age?

Probably not. One of the many functions of

civilization is to make it possible for idiots to survive. In the Stone Age, idiots likely weren't going to survive for very long. They were prey for wild beasts, or simply couldn't feed themselves.

Civilization protects people from their own stupidity. Wild beasts are kept at bay, and smarter people than you have already figured out how to produce a lot of food and build shelters that aren't death traps.

An abundant supply of idiots isn't a warning sign that something is going wrong with civilization, but is instead proof of how effective civilization is at keeping people alive.

Can this drive us to extinction?

Definitely not. Stupid people still possess a well-documented capacity to create more people.

Is there hope?

One way to mitigate growing human stupidity is to build artificial intelligences to pick up the slack. Oh, wait, we've already been over that.

Our best hope is to keep doing what we've been doing. Today, we take it for granted that everyone should have access to a free public education. We take it for granted that a town of any size at all is going to have a library. The results of investing in these things are pretty good!

Thirteen Dooms

And despite the flood of misinformation you can encounter online, we currently live in an age where anyone who desires useful knowledge can easily obtain it. Most of us carry a handheld computer that can pluck books out of thin air.[58]

It's easy to forget just how new this technology is. When I log onto social networks, the level of debate feels a bit like arguments I got into in college. I had little actual knowledge of the world, and would spar with someone else who knew just as little, and our combined ignorance only made us more passionate in defending our opinions.

It took some actual life experience to grow aware of how little I knew, and a smidgeon of humility to accept that I'd need to work the rest of my life to mitigate my own ignorance.

I still hold out hope that the modern internet is in a similar youthful stage. In time, people will realize they have a much higher opinion of their own knowledge than they should, and we'll collectively come to value facts over fiction. We'll develop better tools and systems to process the infinite universe of knowledge before us. I predict a future where humanity obtains a level of wisdom we can't even dream of today.[59]

---

[58] Or, with a click of a button, get a fresh book printed and delivered to your doorstep.

[59] This is, undoubtedly, the most stupid prediction I'll make in this book.

In one dread moment all the sky grew dark,
The hideous rain, the panic, the red rout,
Where love lost love, and all the world might mark
The city overwhelmed, blotted out
Without one cry, so quick oblivion came,
And life passed to the black where all forget

William Wilfred Campbell, *Out of Pompeii*

THE ELEVENTH HORSEMAN: SUPERVOLCANOES!

One of the more uncomfortable realities of the modern world is that we sometimes learn about upcoming world-altering catastrophes that we can do absolutely nothing about. With most dangers, we can think of plausible solutions. We can, potentially, deflect a doomsday comet. We can, potentially, dismantle our nuclear arsenal. Climate change. Famine. Plague. Smart people are working around the clock to stave off these dooms.

And then there's supervolcanoes. Throughout recorded history, volcanoes have created a lot of grief for mankind. Pompeii is a famous example of a town that was wiped off the map[60] by the eruption of Mount Vesuvius in 74 AD. With a population of 20,000, Pompeii was a sizable city for its time, and a thriving hub of commerce and art. The ash that destroyed the city also wound up preserving the art, so that today Pompeii is home to some of the best-preserved murals and sculptures of the ancient world.

It's also home to a more gruesome type of sculpture: Archeologists have preserved casts of over a thousand

---

[60] Figuratively, not literally. It's actually very well mapped.

bodies of those who perished in the eruption. Many of the bodies are twisted into agonized poses, while others poignantly capture mothers still holding their children, or lovers locked in an eternal embrace.[61]

Still, a modern person could look at the tale of Pompeii and find reasons to think they'd probably be able to survive a volcano. For all the destruction wreaked by Vesuvius, the majority of Pompeii's citizens made it out alive.[62] Certainly we're better able to manage the dangers today, with our superior knowledge and technology.

For the most part, this is correct. At any given moment there are several dozen volcanoes erupting around the world, but the annual loss of life seldom equals the death toll Pompeii experienced in a single day. With our seismic tools, volcanic eruptions are forecast long before they happen. With modern transport, whole cities can be evacuated in a relatively short time, and the relative wealth of the modern world ensures that those evacuated usually are provided meals, shelter, and medical care to get them through the crisis. This is yet another terror of the ancient world we've managed to conquer.

Except...

Today's volcanoes are pipsqueaks compared to the true giants of vulcanism, supervolcanoes. The

---

[61] At least, that's the poignant story modern people project on some of the entangled bodies. They could, for all we know, be strangers.

[62] They're all dead now.

volcanoes we typically experience can blow the tops off mountains, throwing debris for dozens of miles, and creating plumes of ash and gas that affect the atmosphere thousands of miles away. But these are small volcanoes.

Their larger cousins are slumbering now, but we know where some are. The United States is home to a famous one: Yellowstone National Park. Its famed geysers and stone formations are evidence of widespread volcanic action beneath the tranquil surface of the park. The most recent eruption of the volcano was 640,000 years ago, and it created a caldera 45 miles across.

How bad would it be if Yellowstone or another supervolcano were to erupt today? It would create conditions similar to the impact of a comet. A Yellowstone eruption could leave wide swaths of North America dealing with falling ash, poison gases, and widespread fires. Then, winter would arrive, and stick around for years.

It's happened before. 74,000 years ago, the Toba supereruption in Indonesia left behind a caldera sixty miles wide and arguably set in motion 1000 years of global cooling. The human population of the era plummeted, and all humans today are descended from a pool estimated to be as small as 3000 individuals that made it through the cold years that followed the eruption.[63]

---

[63] Probably. With events this distant, all interpretations of the evidence are open to debate.

Supervolcanoes are still with us, lurking under multiple continents. Fortunately, while smaller volcanoes can explode rather frequently, supervolcanoes need half a million years or more to build up the pressures required for a supereruption. We might have a hundred thousand years or more before the next one erupts.

But, since we don't have first hand experience with the exact conditions and geological signals that exist before the eruption, we honestly can't be certain of when this particular doom will arrive. It's probably far off. Or, maybe it's next Tuesday.

Should I panic? Will this kill me?

While we don't know when the next supervolcano will erupt, we do know that smaller volcanoes give plenty of warning before they pop their tops. If a supervolcano were building toward an explosion, we'd likely have decades, or even centuries, of geological signals warning us that doom was creeping toward us. There's a very low probability this is going to lead to your death. That said, if there *is* an eruption on the scale of Toba during your lifetimes, the odds that it will kill you are relatively high.

Can this decimate us?

Maybe? It would be difficult to evacuate the population of half of North America. Even if we cleared the land of everyone who lived within 500 miles of the eruption, the possibility of fast death gets swapped for the possibility of a slower death in the killer, years-long winter that will grip the globe.

Can this send us back to the Stone Age?

Probably not. True, a volcano famously brought an end to the Minoan civilization that once thrived on the island of Thera. On the other hand, civilizations not terribly far away, like Egypt, muddled through. A supervolcano would be a true test of resilience for the nations of the world. You could see the social fabric rend and come apart in the face of such a crisis, but it's equally plausible that the difficulties could lead to greater cohesion and cooperation.

Will this lead to extinction?

The close call of the Toba supereruption means we can't completely rule this out. On the other hand, that happened at a time when the totality of human knowledge that had been written down probably matched the totality of all the chimpanzee wisdom that had been written down. A supervolcano isn't going to sneak up on us. We'd have years to prepare, assuming our political leaders took the warnings of the geologists seriously.[64] We'll be fine.

---

[64] Insert your own joke here.

Is there hope?

Let's imagine a worst-case scenario. What if a supervolcano exploded at the same time a doomsday comet slammed into our planet? This seems almost laughably unlikely. Except, *ha ha,* it's happened before. The comet that gouged out the Chicxulub crater happened to hit the planet around the time the Deccan Traps were formed, a geological structure created by a massive, long-lasting eruption. The resulting lava flows are spread over thousands of square miles.

If these tag-team dooms were to occur today, no matter how adaptable humans might be, there aren't many places on Earth where civilization could comfortably endure.

Luckily, Earth isn't the only planet in our solar system. The ultimate survival plan for humanity is redundancy. We need a backup Earth.

The challenge of taking a lifeless, distant world and transforming it into a living biosphere capable of hosting human life is a formidable one. But, there's nothing we've discovered about biology or physics that makes it impossible. We already manage to keep people alive for months at a time on space stations. Moon colonies and Mars bases could be built before the end of this century. The barriers aren't technological as much as they are political. A nation wanting to build such a base would need to devote a lot of money to the project, and the second the project was announced you would predictably get naysayers and grumblers. *Why are they building a city on the moon when they can't even fill the potholes on Main Street?*

The potholes are an immediate problem. Moon bases will require decades of planning and further decades of work to become viable. To transform Mars into a living world is a project of centuries. It's difficult to get people excited about a project that they won't live to see built, and that even their grandkids might not be able to enjoy.

A century is a long time for any human. What if, one day, it wasn't?

*Deck thyself now with majesty and excellency;
and array thyself with glory and beauty.*

Job 40:10

THE TWELFTH HORSEMAN: TRANSFIGURATION!

Surviving as a *Homo sapiens* is sometimes rough, but it's no secret why we endure. Our mighty brains are our greatest asset, allowing us to adapt to all manner of environments. We think our way through most of the challenges the natural world has thrown at us.

But our brains come with a cost. Several costs, actually. For starters, as a species, we're cursed with unusually high maternal mortality, since we're born with unusually large heads. Until relatively recent history, it's estimated that about one birth in fifty would result in the death of the mother. This makes childbirth one of the riskiest of all human activities.

Moms might be at risk during birth, but being a baby was even riskier. In prehistoric times, it's estimated that up to half of all births produced a child that died before reaching puberty.[65] Human babies are both extremely helpless and extremely slow to mature. This is an evolutionary trade off. If we matured in the womb until an age where we were further developed, even more mothers would perish during childbirth. Our horrifying mother/child mortality rates are,

---

[65] The actual percentage is a topic of debate. Prehistoric people were, by definition, terrible record keepers.

paradoxically, proof that large human brains help our species survive.

Another cost of our highly developed brain is a dark mirror of all the risks that accompany birth. From an early age, humans develop an awareness of mortality. Other animals can mourn dead companions, but as far as we know, we're the only animals that can mourn while anticipating death. It's not just the inevitable deaths of loved ones that cause us grief. We can, and do, use a significant portion of our brainpower to contemplate our own mortality.

Of course, a little awareness of death conveys advantages. We're slowly but surely, over a span of decades, whittling away at any number of possible ways to die. We've made progress against pathogens, pollution, and poisons. Fewer and fewer people die on the job. We've cut mortality from transportation to a fraction of what it used to be.[66] We're less likely to be murdered, less likely to drown, and far less likely to perish in a fire than the people of previous centuries.

As a reward for all our hard work, foresight, and precautions… we still die. We can evade every danger, eat right, exercise regularly, swallow every miracle drug, but, in the end, less than 1% of us will celebrate our 100th birthday.

Every birth is a death sentence. Life's a prison with only one exit. Unless this grim verdict has a loophole.

---

[66] You're safer taking an airplane to a destination 5000 miles away than you would be riding a horse that far. Especially if you're going to Hawaii.

Since the dawn of civilization, people have clung to the possibility that they are, perhaps, immortal. This is a basic tenet of most major religions. Perhaps we'll be reincarnated, or we'll live on in paradise, or burn in hell. Whether the outcome is good or bad, our current lives are basically starter projects. Our true home is eternity.

I understand the appeal of such teachings. It can give our brief lives an extra layer of purpose and meaning. The hope of an eternal reward or punishment likely reigns in our worst impulses. There's also a hope, or at least a comfort, in believing we'll be reunited with lost loved ones. The dream of eternal life makes it possible for romantic partners to whisper, *"I will love you forever."*[67]

At the top of the page, I said these ideas have been around since the "dawn of civilization." That's certainly an underestimate. We've excavated Stone Age graves where the dead were buried with tools and pottery. It's plausible to interpret this as proof our distant ancestors already believed in a sort of continuity of life.

However, trusting in beliefs that originated in the Stone Age runs counter to the larger trend of modernity. We're well into our second century where we've made progress by shifting our focus from spiritual concerns and using our powerful brains to better understand matter. We can coax atoms into new combinations to perform nearly miraculous functions.

---

[67] This is certainly sweeter than, "I will love you for an unknown period of time."

Working from the atomic level upward, we can transform raw matter into better batteries, more powerful computers, and more effective drugs. Which raises the obvious question: Why not make use of our mastery over matter to make a better version of *us?*

At its core, human immortality is basically an engineering challenge. There are multiple ways to try to solve the problem.

Few of us die in the bodies we were born in, at least not completely. Our cells are constantly dying off and being replaced with new ones. The turnover is so rapid that our bodies are remade every seven years, though there are holdout organs that don't replace cells as often, such as the heart and the brain.[68]

Aging is basically a function of our cell replication getting increasingly sloppy over time. Scientists think they understand the molecular culprits. The genes in our cells end with a region of repetitive sequences called telomeres. The telomeres exist to help prevent replication errors. When genes get copied, the final portions of the telomere fail to duplicate. The shorter the telomeres become, the more errors there are in cell replication.

So, one path to halting aging would be to find some way to protect or repair our telomeres. Alas, it's far simpler to say this than to do it.

Let's move onto a second option: Why wait for nature to replace all the cells in our organs? We're already

---

[68] Yet another way our brain is a troublemaker.

growing meat in test-tubes. We can clone other mammals, like sheep. There are ethical concerns associated with cloning humans, but in the near future, we might be able to grow new organs outside the body using a person's own stem cells, which can be harvested from body fat.[69] With a fresh heart and lungs every fifty years, and the occasional new liver or kidney, we might be able to extend our lives for decades.

Except... there's that pesky brain again. You can't grow a brain in a test tube and swap it out with the old one and still be the same person. Brains are just as vulnerable to aging as the rest of our body, if not more so. Living into a second century with a healthy body might be a curse if we spend most of that extra time in mental decline.

At present, there's no good way of preserving a person's memories in bulk. There are no magic scanners that can image the cells in your brain, and even if they could, a person's mind changes over time. A brain scan from twenty years ago wouldn't fully capture the person you are today.

Nor are all your thoughts and feelings confined within your gray matter. Your personality is shaped by hormones being released and absorbed all throughout your body. Despite our capacity for reason, we're often driven through life by our guts and our gonads. I'm a techno-optimist, but copying every system needed to back up the memories and emotions of an adult seems

---

[69] A good reason not to diet!

like an engineering challenge we're unlikely to solve. But what if we didn't wait for adulthood?

What if, while we were still in the womb, a chip could be implanted to record all the activity of our brain? This monitor could digitize every word we spoke, encode all the things we see, and even store all that we've touched, tasted, and smelled. This data could become a sort of modern soul, a thing that recalled everything about who we were, yet could still be separated from our body. It wouldn't be our animal self. It would be a thing of pure intellect, a higher self.

If we're sentimental, this higher self could be placed into a freshly assembled clone body grown to mimic our original body at peak condition. Thus, our digital souls might pass from one healthy body to another, finally unlocking eternal life.[70]

But why be constrained by the limits of mere flesh?

We're already building humanoid robots. If we were to pop our digital souls into a machine body, we might leave behind humanity and become superhuman. We could have stronger, tougher bodies, sharper senses, and far more adaptable forms. Freed of the need for air and able to tolerate a wide range of heat, we might comfortably stroll across the valleys of the moon, or leave our footprints on the sands of Mars without bothering to terraform these landscapes.

But why limit our new bodies to human form? We might create new shapes and identities. We could go

---

[70] And a really icky waste disposal problem as we decide what to do with the worn-out bodies.

out among the stars as angels, or as dragons, or as a swarm of tiny gem-winged butterflies. Even more useful? We could exist as an amorphous blob of nanites, and take on any form we desired with only a thought. Man, woman, cat, dog, bird, fish… *anything*. We could experience life from a multitude of perspectives in the course of a single day.

Such a future may never come to pass. But, if it does, the unsettling truth is that the species *Homo sapiens* will finally be extinct. Our advanced robotic forms might gather to watch the last human bodies lowered into their graves. Then, with a shrug, they will turn their backs to all that we've been, and look only toward what will be.

Should I panic? Is this how I'll die?

Nope. If you're old enough to read this book, you're probably too old to ever experience a full digital backup of everything that has gone into making you *you*.

Don't feel bad that you've lost your shot at having your fleshly form transfigured into that of an immortal robot. Immortality would certainly come with its own fresh challenges. Vows of eternal love will be put to an unwelcome test. There might also be a loss of urgency, as the grand tasks you hope to accomplish in life no longer have a literal deadline. Most worrying of all might be the loss of all sense of self. Presumably, your immortal self would understand it was only an echo of a previous self. If it permitted itself to grow and change, would it still be you, or a brand new being? We might spend our endless centuries struggling through one identity crisis after another.

Can this decimate us?

Not any time soon. If the technology were perfected tomorrow, it probably wouldn't be cheap. Digital immortality might be a luxury that only a few people could purchase. But, even if the technology was free, I doubt that it would be embraced. It would be rejected by the devoutly religious as a matter of principle, and baseline skepticism of billionaires and governments would probably keep most people from willingly sticking a chip into their unborn child.

Can this send us back to the Stone Age?

If immortality is perfected, it might send the world a lot further back than the Stone Age. You can imagine future immortals using their powerful technologies to repair the ecosystems damaged by *Homo sapiens*, and clone lost species back to life. When, at last, the world was cleansed of all traces of mankind, the immortals could depart Earth for the last time, and seek out what the rest of the galaxy has to offer.[71]

---

[71] Which contradicts some of the arguments I made in the chapter on alien invasions. Perhaps the aliens who invade us will be our future selves after they've looped through all of time and space and feel ready to come home.

Can this lead to extinction?

In the long term, almost certainly. When humans first walked this planet, we were surrounded by closely related species. Those species are long gone, and we're probably to blame.

We likely killed some of our competition outright. Others, we probably drove to extinction simply because we outcompeted them for resources. Our larger brains made us better hunters, better gatherers, and better able to deal with things like climate change as the Ice Ages ebbed and flowed.

There's no reason to think the pattern wouldn't repeat. Immortal humans in superhuman bodies wouldn't need to grow actively hostile to humanity. They might simply be better at making use of the world's resources than natural humans are. Humans would be fated to go the way of the Neanderthal.[72]

Is there hope?

If the thought of leaving behind our humanity to become a new species worries you, your best course of action would be to actively work to destroy civilization as we know it. If our knowledge and technology keep improving, we'll eventually chart a path to escape from mortality.

---

[72] Unless our replacements keep some of us in zoos.

We can't simply freeze our understanding of the world at its present level. If you can convince humanity to give up on sharing and refining its cumulative knowledge, or convince it to surrender its best practices like, free speech, constitutional rights and global trade, it's possible you might be able to degrade the world enough to carry us into a new dark age, where immortality slips beyond our grasp. We'll still face misery and death, but at least we can die knowing we've retained our humanity.

THE THIRTEENTH HORSEMAN: ENTROPY!

Things fall apart; the centre cannot hold. Yeats knew where this book was heading before I was born. We might dodge peril after peril as a species, and safely find our way through a labyrinth of dooms. We might spread to other planets, and even to other stars, and perhaps, somehow, solve the riddle of eternal life without sacrificing all the things that make us human.

It won't matter. We're doomed. Why? Alas, thermodynamics. It's a fundamental law of the universe that *everything* breaks down.

In the long run, no matter how well we navigate the many dooms ahead of us, the age of humans must one day draw to a close. Of course, that day may be so far distant that no one will even remember what days were.

We experience days because we live on a spinning planet that exposes half its face to our sun at all times. We live through cycles of light and dark. This is something we take for granted as fundamental, even eternal, but our planet's days are numbered.

Our sun is quite stable and long lasting, as stars go, but it's slowly and inevitably burning through its available fuel. As it does so, it's expanding and brightening. On the scale of a human life, this gradual expansion doesn't matter. But, if we did somehow manage to transform ourselves into a race of immortals, we'd have only a few billion years to enjoy the Earth before our oceans boiled away. A few billion years after this, the Earth will cease being a place altogether. The sun will expand so large it will engulf our world, vaporizing it.

Of course, that far in the future, why worry about the loss of a single planet? We'd have long since spread to neighboring stars. Assuming that the speed of light is a hard limit we'll never overcome, traveling between stars may take centuries. But what are mere centuries if we've solved the riddle of immortality?

Outward we spread, making new homes around new stars... but even this must come to an end.

The harsh truth is, the universe is losing stars faster than it's making new ones. The expansion of the universe is constantly pushing new and existing stars beyond the galactic horizon, where the light they emit is so distant it can never reach us. Eventually, the skies will grow darker and darker, as old stars die and others vanish into the unknown and unknowable.

But this far into the future, why worry? Technology will likely be so advanced that we can simply build new stars. Alas, in the long run, we'll simply run out of raw material to work with. A trillion years from now, star formation will end. The expansion of the universe will ensure that the remaining atoms are too scattered through the infinite void to be drawn to one another. (A great deal of matter will also have vanished into black holes during this time.)

The challenge of collecting vastly scattered atoms in the distant future will give way, many trillions of years later, to the difficulty of there simply being no more atoms. All matter is temporary. The fundamental particles are held together by bonds that inevitably break down.

The unstoppable unraveling of our entire universe isn't a flaw in our reality; it *is* our reality. All the forces that lead to certain destruction are the same forces that made our existence possible. If gravity didn't warp and compress matter to the point that atoms are fused together, there would be no stars. If stars didn't burn through their fuels and eventually explode, there would be no elements other than hydrogen and perhaps a little helium. Acts of terrifying, unfathomable destruction, forces powerful enough to tear stars apart, are the ultimate origin, of you, of me, and everything.

But the end is written into the beginning. Everything created is made from something destroyed. Our planet, and everything on it, including us, is built from the ash and dust of long dead stars.

Ashes to ashes. Dust to dust.[73] There's always only been one way this can all end.

---

[73] "In the sweat of your face you shall eat bread, till you return to the ground, for out of it you were taken; for dust you are, and to dust you shall return." Genesis 3:19

Should I panic? Is this how I'll die?

If you live long enough to watch the last stars go dark, count that as a win. But, I feel like I can safely say you won't make it that far. Entropy isn't just chewing up the universe. It's chewing up you. You might escape destruction for a long time, but eventually, something will go wrong. Then, things will get worse. Eventually, something will kill you. You can accept this stoically, or you can panic. Whichever you choose, it won't change a thing.

Can this decimate us?

Didn't you read the first page? Entropy has already decimated the human race. The overwhelming majority of your widely dispersed kin were dead before you ever picked up this book. You may have been told they died of disease, famine, war, or childbirth, but these petty details mask the simpler truth. Entropy claimed them all.

Can this send us back to the Stone Age?

There are crumbling stone walls scattered all over the world, scoured by sands, drowned in shallow seas, and hidden beneath veils of vines. Civilizations rise, then fall, or else they mutate into something new. It feels arrogant to think that our present civilization will defy the longer trend.

Thirteen Dooms

Can this lead to our extinction?

Entropy is the final horseman. It's Death. For now, it's content to reap us with disease and accidents and violence. The grand finale will be a universe full of dying stars scattered through the blackest sky imaginable, with each light fading until nothing remains.

What can we do to stop this?

Nothing. You're going to be alive for a finite time. It might be a long time, but things fall apart. Maybe you have decades before you, or maybe only days. For some of you, reading this book might have burned through your last precious hours[74].

We can't escape the final horseman. It's all over. There's no hope. So what now?

---

[74] Sorry!

"What is the matter with me? I will do something dreadful if I am not careful," she thought, and turning her face to the wall, began trying to force herself to face bravely the fact that many people must live and die alone, even in Winesburg.

Sherwood Anderson, *Winesburg, Ohio*

WELCOME TO THE HUMAN CONDITION

Every end is a beginning. More than 99% of all species that have ever existed are now extinct. Should we mourn their loss? If you like being alive, there's no reason to lament all the death that's come before you.

The elements you're built from might have been forged inside of stars, but the chemicals you're made of are mostly assembled from dead things. You're built of dead meat and dead plants. And don't feel too bad about all the once living things you've needed to devour; at this moment, there's an entire ecosystem busily devouring *you*.

You're covered in mites and microbes. If you dipped yourself in an industrial grade antimicrobial ointment, you'd find the results quite unpleasant. You need these tiny nibblers to keep your skin in good health. The same is true of your gut. With enough antibiotics, you can rid yourself of gut bacteria, then suffer from either diarrhea or constipation until you get these microbes back. You aren't an individual. You're an ecosystem.

As terrible as our own future deaths might seem, civilization is mostly built on the bones of those who came before us. The shapes of the letters you're reading were designed by men who died almost four thousand years ago. The last meal you ate might have included grains first cultivated *ten thousand* years ago.

Farming and the written word are two of the most transformative inventions of all time, yet the names of all the individuals who helped invent and refine them are lost forever[75].

If you can't be remembered for something as important as inventing farming or the alphabet, does anything we do matter at all?

Yes.

Modern civilization couldn't be built in our generation alone. The people who came before us created our present. It's our turn now, to create the future. The greatest thing we can do is offer up our best ideas, put forward the effort of nudging life and order a little further along the path of improvement for a few decades, and then, out of simple politeness, we need to die to give the next batch of people a chance to get to work. No matter how great we've been, we all hit a stage where it's time to step out of the way so others can have a shot at this whole "humanity" thing.

The key to finding beauty in the human condition is to push beyond thinking about yourself. Yes, in the grand scheme, your life is fleeting, and soon forgotten. But the unknown inventors of farming made your time on this planet a little more bearable. You can never be certain, but perhaps something you do will still matter to someone ten thousand years from now, even if

---

[75] To invent the alphabet and not have your name written down is a heck of a raw deal.

they've forgotten your name, even if they never pause to think of how you must have existed.

Our awareness of our certain death, and our understanding that even the memory of us will be erased by time, can leave us feeling lonely.

But, we're not alone.

You're connected to people who survived the eruption of a supervolcano. A million years from now, a child retaining some echo of human DNA might be swimming in the oceans of a terraformed Mars. And you might have helped get them there with the simplest action.

Mundane things you never thought about created this child's future. Maybe your 401k has investments in a company that's building AI, or putting satellites into orbit. Thanks to the interconnected nature of modern commerce, a small fraction of your wages are financing technologies you've never even heard of.

Forget about money. Maybe all you did was take the time to sort your plastics before recycling them. The oceans of Mars could be modeled on the trash-free, unpolluted seas of a future earth that you helped make possible.

If you want to feel like your life matters, remember that, every day, you're crafting the future, even if you won't be around to see it.

We're doomed. Accept this with a little dignity and grace. Hold onto the hope that, long after you and I are gone, someone—or *something*—will step into our footsteps to appreciate all the beautiful tomorrows.

AFTERWORD
THE RESERVE HORSEMEN

I made a footnote in the second chapter mentioning that I could name 30 potential dooms. This wasn't just a throwaway joke. There are a lot of potential doomsdays I considered while assembling this book.

Some, like vacuum decay, fall far beyond my somewhat limited powers of explanation. Others felt a little closely related to topics I was already covering. After tackling comets and asteroids, does it really add any value to mention that there are also rogue planets that might doom us?

Still, I'd hate for you to have paid good money for this book only to be completely blind-sided by a doom I knew about that I forgot to mention. So, here are seventeen more dooms you might want to look into. Like I haven't given you enough to worry about already.

1. **Ice Age:** Ice ages ebb and flow. We're in an interglacial phase, but the ice sheets will be back.
2. **Gray goo:** Nanotech might run amok.
3. **Vacuum decay:** Just google this one.
4. **Pole reversal:** It's happened before. It will happen again. Probably a very bad day for our tech.
5. **Solar flare:** Another tech fryer.

6. **Supernovas:** We don't need to be orbiting a star for its explosion to destroy us.
7. **Sterility**: Something seems to be a little off with our sperm.
8. **Rapture:** That whole Jesus thing. A lot of people believe this is the big one!
9. **The "Big One:"** The other big one is California literally sliding into the ocean and the tsunamis and other bad things that follow.
10. **Insect apocalypse:** Pesticides, habitat loss, and climate shift has led to a reduction in insect populations. Could the whole food chain collapse?
11. ***Homo superior***: Mutation and evolution are likely working on our replacement.
12. **Black holes:** Maybe from outer space. Or maybe we'll accidentally create one.
13. **Information poisoning:** For most of history, technology helped us become collectively smarter. But, in the age of social media, fake "facts" and misinformation often spread faster and further than truth. Will a flood of falsehoods strip us of the capacity to deal with world-ending problems?
14. **Rogue planets:** The galaxy is full of them. One wouldn't have to hit us directly to create any number of catastrophes, including shifting our planet into an altered orbit.
15. **Inequality:** Capitalism doesn't create poverty, but it does create extreme disparities. The resulting political distortions have led to world wars and the collapse of nations.

16. **Technological collapse:** Network failures are so mundane we just shrug and wait if we're unable to log into our email or take a peek at our checking account. We trust that our services will come back online soon. But what if they don't?
17. **The simulation crashes:** And speaking of networks, some people make a weirdly good case that we're living inside one. What happens when the heavenly janitor accidently trips over our power cord?

Okay. That's enough doom and gloom! Put the book down! Go outside and enjoy the sun, or maybe the stars, or clouds, or the pale luminous haze, depending on the time of day, and your local weather conditions.

Whatever sky you're under, cherish it. It won't be there forever, you know.

ABOUT THE AUTHOR

James Maxey's mother warned him that reading too many comic books would warp his brain. She was right! Unsuited for decent work, Maxey now ekes out a living writing down his demented fantasies about superheroes, dragons, and monkeys. He's written an absurd number of books, including *Bitterwood, Greatshadow, Nobody Gets the Girl,* and *There is No Wheel.* He's also been awarded the title of Piedmont Laureate, which is evidence of some talent for either blackmail or bribery. If you encounter Maxey in the wild, don't panic! He's mostly harmless.

For more information, visit www.jamesmaxey.net!

BOOKS BY JAMES MAXEY

The Dragon Apocalypse:
Greatshadow, Hush, Witchbreaker, Cinder
Bad girls! Big dragons! After a failed quest to slay the dragon Greatshadow, Infidel (the heretic), Sorrow (the witch), Gale Roamer (the pirate captain), and the Black Swan (the crime queen), must team up to save the world from wrathful dragons!

The Bitterwood Saga:
An epic war between dragons and mankind! When the mysterious archer known as Bitterwood kills the son of the dragon-king, it kindles all-out war between human rebels and their dragon overlords. The adventures continue in the novel *Dragonforge* and *Dragonseed*, plus secrets are revealed as the origins of Atlantis are explored in the prequel novel *Dawn of Dragons!*

The Dragonsgate Trilogy: Devils, Spirits, & Angels
The spinoff series from the world of Bitterwood! Graxen and Nadala are sky-dragons exiled beyond the Cursed Mountains. When they find a mysterious machine in the ruins of Oak Ridge Tennessee, a hole is ripped in the fabric of space and time. This begins a quest across parallel universes to save the Earth from the planet devouring Waste-Wyrm!

The Nobody Series:
Nobody Gets the Girl, Burn Baby Burn, Covenant
Superheroes and supervillains battle for supremacy, flattening whole cities. Who can save us? Nobody!

The Lawless Series: Cut Up Girl, Big Ape, Victory
Golden Victory! She-Devil! Atomahawk! Smash Lass! The greatest heroes of mankind join forces as the Lawful Legion! This... isn't their story. Lawless follows the superheroes who haven't been invited to the team, outcasts like Cut Up Girl and Big Ape, with lousy powers and worse attitudes. But in a world of alien invaders, time tyrants, and genius dinosaurs, even losers sometimes have a shot at saving the day.

Bad Wizard
Return to Oz! Dorothy Gale, reporter, is investigating Oscar Diggs, Secretary of War. Only she knows that Diggs was once the Wizard of Oz. Now that Diggs has constructed a fleet of armed zeppelins, can Dorothy prove to her editor that America is getting ready to go to war with a cloud island ruled by witches?

Short Story Collections:
There is No Wheel, The Jagged Gate, Life in a Moment
Circus freaks, junkies, witches and monkeys populate these offbeat tales. Write what you know, they say.

Anthologies edited with Cheryl Maxey:
Beware the Bugs! Rockets & Robots, Paradoxical Pets.
James and Cheryl assemble an international cast of authors to present exciting science fiction and fantasy short stories suitable for middle grade readers.

Non-Fiction: *Write! Daydream, Type, Profit, Repeat!* and ***Crytpids: How We Know They are Real.*** The titles pretty much explain these!

LINKS

Review this book!
Follow the QR code to the Amazon listing for this title.

Read more by James Maxey!
Follow the QR code to his Amazon Author page.

Made in the USA
Columbia, SC
01 February 2025

468accc9-16df-4886-8e7d-bcf4039afde6R01